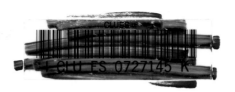

People and Woods
in
Scotland

PEOPLE AND WOODS
IN
SCOTLAND

A HISTORY

Edited by

T. C. Smout

EDINBURGH
University Press

Edinburgh University Press Ltd
22, George Square, Edinburgh

Typeset in Minion
by Pioneer Associates, Perthshire, and
printed and bound in Great Britain by
The Cromwell Press, Trowbridge, Wilts

A CIP record for this book is available from
the British Library

ISBN 0 7486 1700 0 (hardback)
ISBN 0 7486 1701 9 (paperback)

Forestry Commission

Contents

Acknowledgements

Our first debt is to the Forestry Commission who commissioned this book, and in particular to Marcus Sangster, whose initiative suggested it in the first place. Tim Yarnell and Peter Quelch of the Commission were also of great help, and points in the text were illuminated by Barbara Crawford, David Alderman and Magnus Magnusson. Robert Crawford and Derick Thomson kindly gave permission to reproduce their translations of poetry quoted in the Introduction. Highland Birchwoods provided additional financial support for Chapter 8. John Davey smoothed the way to publication with Edinburgh University Press, to whose efficient production staff we are also most grateful. Margaret Richards, with infinite patience, care and accustomed good humour, processed the text. The staff of St Andrews University Library and Photographic Service were extremely helpful.

All the illustrations and figures are individually acknowledged throughout the book, and we are most grateful to all concerned. Particular debts, however, are due to the staff of the Royal Commission on the Ancient and Historical Monuments of Scotland, of the National Museums of Scotland, of the National Archives of Scotland, of the National Library of Scotland, of the Highland Folk Museum, of Perth Museum, of Aberdeen Museums and Art Galleries, of Elgin Museum, of the Woodland Trust and of the Forestry Commission picture library. We thank them all most sincerely.

CHRIS SMOUT
Centre for Environmental History and Policy
University of St Andrews

List of Plates

List of Figures

List of Colour Plates

To be found between pages 148 and 149.

Trees in Scottish Life

CHRIS SMOUT

When the cold of the last Ice Age finally began to withdraw eleven and a half millennia ago, Scotland was left a tundra bereft both of trees and people. Gradually, over the next four millennia, first birch and hazel, next pine, oak, elm and ultimately alder clothed the land. The first people, hunter-gatherers, left traces on the shores of Rum about nine thousand years ago, and then up and down Scotland. People and trees were colonists together: we have a long association.

Woods were at first considerably more successful than people in establishing their dominance in the landscape. Trying to estimate what proportion of Scotland's surface was covered with trees at the time of the woodland maximum about five thousand years ago is not easy, but it is probably true to say one half or more.

Thereafter, woodland began its long retreat, driven back by a combination of climate change and human interference. The congenial summer warmth and winter cold of the earlier prehistoric period gradually and irregularly gave way to the sort of oceanic climate we know today, with cooler wet summers and milder winters, less favourable to tree growth but encouraging to the creeping spread of peat bogs that made a poor nursery for tree seeds. In the Neolithic, human beings learned farming and their numbers multiplied to match their new skills. To run their herds and plant their seeds, they first used natural open spaces in the forests, then cleared forests for their farms. Moor or bog might ultimately take over, for not all prehistoric farming was sustainable or maintained in the same place. By the time of the late Iron Age, however, in the centuries immediately preceding Roman attack, farming had already cleared much of the south of Scotland of trees, while in the north the spread of bog and heath assisted by the nibbling of stock had also considerably reduced the original cover.

The Romans were only briefly present and probably made little difference. The tide of trees ebbed and flowed in the following centuries, but the trend was remorselessly downwards until, on the eve of the Industrial Revolution, only about 4 per cent of the land remained covered with wood, one of the lowest percentages in Europe, comparable only with Ireland or Iceland. It remained at this low ebb until after 1900. Then, in the twentieth century, an astonishingly successful movement to reclothe the land with trees was led by the Forestry Commission. It brought the area under wood up to its present level of about 19 per cent of the land surface excluding the Western and Northern Isles, 17 per cent if they are included. The trees planted, however, were not the old natives but introductions from North America, where they had been discovered by the great Scottish plant-collectors of the nineteenth century, above all Sitka spruce and Lodgepole pine so well fitted for commercial production. Only at the very end of our story did there arise another movement, sometimes opposed to the first but always ready to seek accommodation. This time it was to preserve, respect and extend the fragments of our native woods, and it, too, aroused the passion and commitment of foresters. Today, the land is again well covered with wood, and likely to become still more so as our new century progresses. Woodland is managed now for many purposes – for timber production, for nature conservation, to provide refreshment for the human spirit, and by-and-large they complement each other excellently. We can be proud that it has become so.

The relationship between people and trees in Scotland has been a deep and important one at many levels, but not always the same. Archaeologists, concerned above all with material remains, emphasise the extraordinary usefulness of timber to people in their homes. For all the millennia of our most distant past, and for most of the centuries of our more recent past, wood in most places was the main building material, certainly in terms of keeping the house up. When stone houses became characteristic of Scotland – in different places between the sixteenth and nineteenth centuries – timber was still needed for the roof, the ceilings, the wainscot, the floors, the panellings. We might not grow it (more and more was imported after 1500) but we had to have it. In rural Scotland (and most people lived in the countryside until well into the nineteenth century) wooden material was also needed for fencing and for tool-making, and the use of the woods themselves for sheltering and grazing stock was often crucial for winter survival.

In this situation, how did it happen that the amount of land under wood had fallen to such extremely low levels by the eighteenth century? There are probably three main reasons. First, the use of timber for fuel was not important in Scotland, given its abundant supplies of peat and coal. Almost everywhere, it was a more efficient use of human energy to turn to these fuels rather than to wood, so a major incentive towards the protection of woodland in some other European countries (for example, France or Denmark) was lacking in Scotland. Where wood was used for fuel, as in the charcoal iron industry of Argyll, its use actually encouraged the protection and wise management of the resource.

Second, the ease with which Scottish towns along the east coast could import cheap and excellent timber from Scandinavia and the Baltic obviated much of the need to grow supplies within Scotland. The burghs of the Firth of Forth in the seventeenth century could sail to Norway within two days and load up with excellent pine from woods that came down to the sea. Why sail to the Moray Firth, which would take just as long, to wait on uncertain and inferior supplies floated down the Spey? In the nineteenth century, the timber of North America was delivered cheaply on the quays of Glasgow, often in return for cargoes of emigrants going in the opposite direction, and the new railways took it the length and breadth of the country. Even the late Victorian owner of Rothiemurchus forest had his new house panelled in Canadian softwood. When wood pulp then came to be used for paper, it came into the ports in the same way.

Third, woods could decay and fall not because they were not useful, but because the mounting pressure of beast and humans and the subtle but ultimately serious alterations in the climate gradually prevented their regeneration. One may lament, for example, the folly of the inhabitants of Strathnaver around 1800, whose misuse apparently led to the decay of the woods important for over-wintering cattle, because they could not, or would not, lessen the pressure of their animals to allow the regrowth of the trees. But an early twenty-first-century world that cannot, or will not, lessen the pressure of carbon-dioxide emissions in the face of global warming is in no position to pass harsh judgements on the Sutherland peasantry of 200 years ago.

In the twentieth century, it was the reality of war that brought the trees back. The first submarine blockade led to the foundation of the Forestry Commission immediately afterwards – how could the country strategically survive another struggle without wood? The second, in which just the need to supply 300 Scottish coal mines with pit-props almost exhausted native wood supplies, underlined the lesson. Shortly after the Second World War, however, it became clear that any future conflict would be over in a few nuclear flashes, and the strategic imperative evaporated. By then, however, there was much water of government help under the keel of the Forestry Commission and private woodland owners. The industry sailed on, not always perturbed by the abstract economic reality that timber could usually be grown abroad and marketed in Britain very much cheaper than timber could be grown and marketed in Britain. Home-grown wood still formed (and forms) the basis of a substantial rural industry.

That this should be so is due in large measure to the fact that trees have always meant more to people than just their economic value. Utility was, and remains, important. But they have also had cultural and spiritual meaning, changing in differ-ent periods, and much more difficult for the historian to uncover. We know from Roman sources and from archaeology that British Celts shared with their Continental counterparts a religion in which sacrifices of wealth, animals and human beings were made in natural places: bogs, lakes, wells and sacred groves. Though none of the written evidence relating to Druids and human sacrifice is specific to Scotland, the discovery of the eerie and powerful wooden image from Ballachulish lends strength to the notion that there were great Celtic gods of the forest here, as well.

It was a technique of the Christian missionaries, sanctioned by the papacy, to ease conversion to the new faith by absorbing something of the old, including the cele-bration of Christmas close to the winter solstice. In this context, it is fascinating and moving to read the great eighth-century poem about the cross on which Christ was crucified, 'The Dream of the Rood', written in Northumbrian Anglo-Saxon, from which the Scots tongue itself is descended. Part of the poem is carved on Ruthwell Cross in Dumfriesshire. This extract from the poem is from the translation by Robert Crawford in the *New Penguin Book of Scottish Verse*:

> I saw glory's tree
> Shawled with light, joyfully shining;

> Gems had clothed those forest branches,
> Yet through gold's glint I still could glimpse,
> Tokens of torture, of that first time
> Blood ran from its right side.

The tree describes its terrible experiences as the instrument of God's death, and the poet ends:

> Now, more than all, I live my life
> Hoping to see that tree of glory
> And worship it well.

It is not difficult to see how worship of the sacred trees associated with human offerings and worship of the cross associated with Christ's sacrifice could meld into one another.

In Gothic architecture, the reference to trees in foliate capitals and arches becomes easy to understand. Even the mysterious figure of the 'green man' carved in churches, on roof bosses and choir stalls appears as some kind of symbol of humankind's connection with nature. Exactly what it means is hard to determine, but the basic feature is always a mature male head with foliage issuing from its mouth and surrounding the face. Sometimes, as at Melrose Abbey, it is a death's head with leaves coming from the eye-sockets, sometimes, as among the hundred or more compelling green men at Rosslyn Chapel it is (see Plate 1) a face so alive as to seem a portrait of an actual person. Common throughout western Europe, the green man appears to speak of the power of nature, of the force of creation, of the cycles of life and death, and of resurrection. All flesh is grass, but new life springs with the leaves, as it does every year in the wood.

With the Protestant Reformation of the Lowlands in 1560 came a total fracture of this kind of symbolic portrayal of trees and foliage, but in Highland Gaelic society there survived a closer sense of continuity with a very old past. Some of the Irish Gaelic law codes of the first millennium AD list very specific hierarchies and characters of trees – oak, hazel, holly, ash, yew, pine and apple being the nobles of the wood; alder, willow, hawthorn, birch, elm and gean the commoners; aspen, juniper and so forth the serfs and slaves. If this seems at first to be something from another country and long ago, that is to underrate the longevity and unity of Gaelic culture between Ireland and Scotland. This becomes surprisingly clear in a seventeenth-century poem by Sìleas na Ceapaich, also from the *New Penguin Book of Scottish Verse* translated by Derick Thomson. It is a lament for Alasdair of Glengarry by his mistress:

> The yew above every wood
> the oak, steadfast and strong
> you were the holly, the blackthorn,
> the apple rough-barked in bloom;
> you had no twig of the aspen,

PLATE 1 *A green man in Rosslyn Chapel, Midlothian. There are over one hundred representations of this enigmatic medieval symbol, carved in different ways, within the chapel. (Antonia Reeve, and Rosslyn Chapel Trust.)*

> the alder made no claim on you,
> there was none of the elm-tree in you,
> you were the darling of lovely dames.

To this day a residue of the belief in the magic and moral qualities of trees survives in the half-remembered notion that a rowan planted outside a home will keep witches from the door. In the more distant past, no tree had deeper cultural resonance than the ash. Norse mythology told of Yggdrasil, the great ash tree, Odin's horse, that united hell, earth and heaven. Its roots were in the underworld, gnawed at by a dragon and seven snakes, its trunk was on earth and its crown held up the sky. The Sinclair Earls of Orkney built in their church of Rosslyn Chapel a Christian representation of Yggdrasil, famous now as the 'Apprentice Pillar' (see Plate 2). At its base, in the underworld, eight dragons chew on the roots. The leaves of a vine, always an emblem of Christ, twist around the trunk of the pillar, embracing our life on earth. Grapes adorn the capital, as the life of Christ came to fruition at the gates of heaven. So Christ and Odin come together at Rosslyn just as the cross and the sacred grove came together in the *Dream of the Rood*.

PLATE 2 *The Apprentice Pillar in Rosslyn Chapel. The pillar represents Yggdrasil, the mythic ash tree of Norse legend, with eight dragons gnawing its roots. The Christian vine climbs up it and fruits at the top. (Antonia Reeve, and Rosslyn Chapel Trust.)*

Norse ideas also suffused the world of the western Gaels, and Thomas Pennant in 1772 described the practice of midwives giving new-born babies 'sap from a green stick of ash held in the fire so that the juice oozed out onto a spoon held at the other end', capturing the tradition of ash as the tree of life. British forces in 1746, as part of their campaign of terror after Culloden, deliberately burnt an enormous ash, 58 feet in circumference, in the churchyard at Kilmallie, the parish church of the Jacobite chief Cameron of Lochiel. The tree was said to have been 'held in reverence by Lochiel and his kindred and clan, for many generations', and from its girth could have been a relic of pre-Christian worship.

Large and ancient trees were often associated with heroes or villains of the historical past, real or imagined. There were at least three Wallace's oaks where he is supposed to have hidden from his enemies. One in the Torwood, Stirlingshire, was always reserved from the axe when the surrounding trees were sold, until eventually it succumbed late in the eighteenth century to old age and to vandalism from visiting tourists seeking momentoes. The other two were at Elderslie, Renfrewshire (see Plate 3), and Methven, Perthshire. Then also in Perthshire there was Malloch's oak, named after a profiteering grain merchant said to have been strung from its branches in a famine, the Birnam oak by Dunkeld (see Colour Plate 8), supposed to be the last remnant of Shakespeare's moveable wood in Macbeth, an old thorn at Port of Menteith used in lieu of the burgh cross, and of course, the Fortingall Yew in Glen Lyon, sometimes described as the oldest vegetation in Europe, where the gullible are still occasionally told that Pontius Pilate played among its branches as a boy (see Plate 4).

Such trees of great antiquity are deeply moving to the human spirit. We respond to this day to ancient, twisted Caledonian pines (some already old in Jacobite times), to gnarled oaks in the medieval hunting park at Cadzow and to the strange pollarded alders with rowans atop them that grow on the hillsides of Glen Finglas or Strath Gartney. Measuring trees that were particularly old, or particularly tall, or particularly wide, first became a fascination to eighteenth-century gentlemen and continues to this day, as exemplified by the work of the Tree Register of the British Isles.

Perhaps Scotland gradually felt emotionally starved by the slowly withering symbolic link between people and trees. The Scottish people certainly took readily in the twentieth century to a revival in the form of the Anglo-German festive and seasonal symbolism of Christmas tree and mistletoe. Ironically, the trees and berries are invariably of non-native species, Sitka spruce, Norway spruce, grand fir and the mistletoe itself, and those who buy them in shops and supermarkets seldom envisage them as growing wild. Never mind, the greenery reconnects us with something ancient.

It is sometimes said that people before the eighteenth century found woods scary, associating them with frightening spirits and menacing animals. If they did so, it left little trace in Scotland, possibly because the woods in historical times were already so limited in scale and so thoroughly under human management. On the contrary, sixteenth- and seventeenth-century writers are all very positive: for example, to Bishop Lesley in 1578, Easter Ross is 'mervellous delectable in fair forrests, in thik wodis', and to Gordon of Straloch in the next century a hill near Ballater was 'occupied by a beautiful forest of tall evergreen firs of immense size'.

PLATE 3 *Wallace's Oak, Elderslie, Renfrewshire, now disappeared. It was already dying back when drawn in 1826 by J. G. Strutt, but was 21 feet in circumference at the base and still 67 feet high. Legend held that Wallace and 300 men hid in its branches. (St Andrews University Library.)*

By the end of the eighteenth century, the new arbiters of taste, the romantics, were not only enthusiasts for trees, but close observers of their form and character. Thus William Gilpin, keeper of the canon of the picturesque:

Remark the form, the foliage of each tree
And what its leading feature. View the oak,
Its massy limbs, its majesty of shade;
The pendent birch, the beech of many a stem,
The lighter ash, and all their changeful hues
In spring or autumn, russet, green or grey.

Both Wordsworth and Burns leapt poetically to the defence of ancient trees threatened by the axe, and by the nineteenth century some were expressing a real dislike of introduced trees, even larch, which Lord Cockburn considered 'stiffened and blackened the land', just as others were enthusiastically competing to decorate their estates with the latest American find.

Then there is the question of woods as places for people actually to be in, as well as to admire. A visitor to the past would be impressed by how populated they were, by swineherds in the Middle Ages running their pigs among the acorns, by children guarding other animals, cattle, sheep and horses, in the wood pastures, by charcoal burners and tan-barkers making fuel and tanning materials for industrial use, by

PLATE 4 *The Fortingall Yew as it was in 1826. Estimates of its age range up to 3,000 years: it must in any case long predate the original church at Fortingall. Mourners are carrying a coffin through the hollow trunk to its final resting place. From J. G. Strutt,* Sylva Britannica *(1826). (St Andrews University Library.)*

those who gleaned nuts and fruit or made 'candle-fir', and of course by people felling, coppicing and pollarding to get the timber they needed for their homes and tools. Burns's lovely poem 'The Birks of Aberfeldy' celebrates woods as quiet places for courting, but couples would need to choose their moment if they were not to be stumbled across by their gossipy neighbours. In the poet's own day things changed, as landowners tried to keep animals out of the woods in the interest of better woodland management and then, as the price of native timber collapsed in the next century, owners increasingly reserved the woods for shooting and personal pleasure. Woodcraft declined as the link between farming and woods was cut, and to people who lived in the new industrial cities woods were simply part of the closed, privileged world of the countryside.

The business of hunting, of course, had always been a matter of status and privilege, as is clear from the depiction in early times of fine horseman out on the hunt with their retainers and hounds on Pictish and medieval stones (see Plate 5). It was made even more explicit by the forest laws of the Middle Ages. By the nineteenth century, stalking the red deer was a solitary craft of the open hill, but it had only recently become so. 'No trees, no deer,' said the factor of Glen Finglas to the Earl of Moray around 1710, and the usual form of hunting then was to drive the animals through the wood into an enclosure where they could be massacred. By Victorian times sport in the woods was mainly shooting roe deer, pheasants and woodcock, as it has remained to this day, though with an increasing emphasis on the managed pheasant shoot. Poaching of course occurred, as it had since the Middle Ages. In many places and for a long time it was considered no great offence, but, beginning at around 1780, game laws of a discouraging ferocity sought to prevent it. Increasingly, though never everywhere or all the time, only owners and their guests were welcome in woods.

This makes all the more remarkable that in the course of the twentieth century the population of Scotland not only rediscovered the woods, but came to use them in extraordinary numbers, so that today about two million Scots visit our woods and forests a year. This was very largely due to the work of the Forestry Commission, who practically from the outset took the view that the nation's forests should be open to the people. A milestone was the creation in 1936 of the Argyll Forest Park, where the public was encouraged to walk in the woods and on unplantable land, followed by other parks at Glentrool, Glenmore, in the Trossachs and on the Borders. For Roy Robinson, the Forestry Commission chairman from 1932 to 1952, provision of facilities for walkers and families in as many ordinary forests as possible was a personal passion. In 1944 he exclaimed to the commissioners that John Dower, author of the report that was to lead to the establishment of National Parks in England and Wales, had got it all wrong: 'Mr Dower had entirely missed the point that there was much more of interest in walking through a wood than over a bare hillside.' In the event, the concern of the Commission with access, although it much troubled the Treasury, who considered it a diversion from the business of growing trees, helped in the 1980s to save it from privatisation. The suspicion of so many urban voters in marginal constituencies that private forestry companies would not allow recreation in the

PLATE 5 *The grave slab of Murdoch Macduffie, 1539, in Oronsay Priory. The birlinn at the foot, so like a Viking ship, represents mastery at sea. The deer represent possession of land and aristocratic pleasures. The great sword is flanked by fabulous beasts and foliage. (Royal Commission on the Ancient and Historical Monuments of Scotland: Crown copyright.)*

woods in the same way as if they were nationally owned made a marked impression on ministers.

By then, however, an increasing number of private landowners had already begun to follow the lead of the Buccleuch estates and others in opening their woods to the people, and the Woodland Trust had been founded with the aim of buying and creating native woodlands fully available to the people.

Other initiatives followed. Reforesting Scotland took the lead in presenting the social, economic and ethical case for reclothing the land with native woodland. Highland Birchwoods took steps to encourage the use of birch, and to increase the understanding of native woodland more generally. Trees for Life and Scottish Native Woods were prominent in practical actions locally. The Royal Scottish Forestry Society began planting a large native woodland at Cashel on Lochlomondside. Some of these and many other community woodland schemes were supported by the Millennium Forest for Scotland Trust. The Central Scotland Countryside Trust has begun to plant on that bleak land, often scarred with past industrial dereliction, between Edinburgh and Glasgow and the Borders Forest Trust has ambitious plans to restore a 'wildwood' to land at Carrifran, where there has been no forest for a very long time. The state itself introduced farm woodland premiums. By the start of the present century there was a plethora of initiatives intended to create opportunities for people to use, understand and enjoy, at every level, the revived woods of Scotland, especially but not only the native woods.

So, today, people are reconnecting with trees, and forestry in Scotland is multipurpose: it brings delight and has use. This book is multi-authored. Thirteen of us

PLATE 6 *Designing with trees. W. S. Gilpin contrasts bad landscape design (top) with more natural and picturesque design (foot). The same principles apply to this day. From* Practical Hints upon Landscape Gardening *(1832). (Christopher Dingwall.)*

have come together to tell the story from the retreat of the ice to the prospects for the future. It calls on an expert in pollen analysis to examine the most ancient patterns of woodland distribution, on archaeologists to describe how wood was put to good purpose (especially in buildings), on historians and foresters to explain what we know of woodland management and the social contexts of his use and enjoyment of the ages, on ecologists to comment on changes humans produced in the wood, on a geographer to examine the crystal ball. It deals with the subject chronologically. Its especial focus is the native trees that comprised most of our woods until 100 years ago, but it also deals with the new species that were used first to decorate the landscape, then as the basis of an industry in which the Forestry Commission took a remarkable lead.

It has been an absorbing story to write and to edit: we hope it will be just as enjoyable and useful to read. At the end of the book a bibliographical section recommends further reading. Those who need additional information on the academic sources for the text are welcome to contact the editor at the Centre for Environmental History and Policy, University of St Andrews, KY16 9QW. Readers may also wish to know about the Scottish Woodland History Discussion Group (c/o The Secretary, Centre for Environmental History and Policy, University of Stirling, FK9 4LA), which holds annual meetings, where everyone is welcome with an interest in the past of Scotland's forests and woods.

CHAPTER ONE

Living in the Past: Woods and People in Prehistory to 1000 BC

RICHARD TIPPING

INTRODUCTION

We live in a prolonged warm period called an interglacial. As the word implies, this is a time between successive glaciations that have had global impacts, and which affected in very profound ways the country we now call Scotland, not least because the Highlands repeatedly nourished large ice-sheets and valley glaciers. The last ice in Scotland rapidly melted some 11,500 years ago. As the climate warmed, diverse landscapes within Scotland developed: soils matured, plants colonised and changed from grasslands to uniform shrub-covered slopes, and trees eventually spread to every part of the country, supplanting in turn the shrubs.

People probably moved into Scotland with the trees – perhaps because in wooded landscapes were the large mammals that were their principal sources of food. For almost half the time since the last glaciers, people in Scotland have hunted animals, collected plant foods and fished in rivers and lochs and on the coast. The woodlands evolved with people as explicit elements of ecological processes that we too often think of as 'natural'. Indeed, there was probably no 'natural' woodland if we think of nature independent of human beings. Some 6,000 years ago, people began to move from a hunter-gathering-fishing economy to one based on domesticated animals and crops, but the change to agriculture may have been more cautious than we once thought. For a long time, until perhaps the Iron Age 2,500 years ago, it is likely that people's engagement with woods was not deleterious to trees or lastingly damaging.

It would be inappropriate to think that our woodlands were controlled by people, however. There are a myriad different destabilising factors that also drove woodland change, ranging over all temporal and spatial scales, from the death of individual trees to climatic shifts that affected whole regions. A key observation is that disturbance is a constant feature in woodlands, however we might think of a stroll in our local woods as the epitome of tranquillity and calm. Just as we now see disturbance as funda-mental, we have also begun to understand that woodlands do not represent end-points in landscape evolution, even though individual trees are long-lived. We used to think that plant communities developed in the post-glacial period to their fullest expression in woodland, and that woods the same as we see today were 'climaxes' in landscape development. This view brings with it the unconscious assertion that there is a purpose to long-term vegetation change. It also introduces the idea that woodlands have more of a 'right' to grow than other ecosystems, like heather moor, for example, and this can influence approaches to conservation. But a perspective which considers woodlands changing over thousands of years shows that our present woodlands are assemblages of different species that are essentially ephemeral. In parts of Highland Scotland such as the 'flow country', trees are nowadays probably not 'natural' components of the landscape at all.

This chapter will attempt to describe the development of our woodlands over the time from the last ice age, 11,500 years ago until around 2,500 years ago when human destruction of this resource became, in parts of Scotland, large scale, extensive and transfiguring. The chapter will stress the diversity and constant change in these post-glacial landscapes. It will try to describe how woodland 'loss' has occurred in different ways over this period, but will also show that the reasons for its disappearance are varied and that human beings are not always culpable. The chapter will also be an exploration of what it was like to be in these earliest woodlands, to move through and be within this most valuable resource.

Before the Trees Arrived:
Scotland after the Last Ice-Sheet

The present interglacial, the Holocene period, probably had two beginnings, or rather it had what we now see as a false start. Around 14,500 years ago the ice melted from huge ice-sheets that covered most of north-west Europe. In Scotland temperatures soared to equal those of the last few thousand years, and microbes, mosses and lichens allowed soils to develop on bare rock surfaces and glacial debris. Over some two thousand years grasslands spread, and tall shrubs like juniper and willow, and later tree birch, colonised parts of southern and eastern Scotland (see Plate 1.1). These areas have been described as looking like open parklands, although we have few measures of how dense the canopy of birch trees was. This was in every way the start of an interglacial, but by 12,500 years ago the birches had gone, and species-poor grasslands, sedge tundra and crowberry heath dominated. Organic soils began to deteriorate, and to erode through freeze–thaw processes into lake basins and rivers as the interglacial was halted. Eventually glaciers once more formed as a large ice-sheet in the Grampians, centred on Rannoch Moor but flowing in radiating broad valley glaciers: elsewhere in Scotland glaciers reoccupied the high corries (see Fig. 1.1). Away from the glaciers, the country resembled today's high arctic, empty save for herds of reindeer and giant deer, and carnivores like bear, wolf and lynx that tracked them. Human beings were, as far as we can detect, not a part of this bleak landscape.

This astonishing climatic collapse persisted for perhaps 800 to 1,000 years. It was caused by the abrupt cessation of warmth-bearing North Atlantic Ocean currents like the Gulf Stream, though what caused the breakdown of this enormous ocean circulation system is unclear. Computer models show, however, how comparatively easy it is to put a stop to such grand global climate changes. Further, climate scientists

PLATE 1.1 *Juniper grows as very tall shrubs, almost trees, on calcareous soils in upper Teesdale, northern England. Extensive stands like these probably grew immediately after deglaciation around 14,000 years ago, before birch colonised and outcompeted juniper, but were fragmented by the glacial climates that followed. Juniper then had to re-establish, and again for a short time around 11,500 years ago dominated the landscape. (Forestry Commission.)*

FIGURE 1.1 *A reconstruction by Dougie Benn (University of St Andrews) of the ice-sheet and valley glaciers that filled the Cuillin Hills of Skye 12,000–11,500 years ago. Larger ice-sheets drained the western Highlands, while smaller corrie glaciers were typical of areas further north, east and south. These huge glaciers melted abruptly at the start of the Holocene interglacial. (From C. K. Ballantyne (1989),* Journal of Quaternary Science, *4 (2), pp. 95–118 with permission.)*

have begun to show that comparable changes may have occurred within the present interglacial, and are capable of happening in the near future. The second beginning of the interglacial, 11,500 years ago, was astonishingly abrupt. Estimates from the annual layers of ice-cores from Greenland show that glacial climates were replaced by fully interglacial conditions over less than a century, probably over 5–10 years. This rapid amelioration left across Scotland a 'clean sheet', where most soils had been set back to bare unvegetated surfaces, where shrubs like arctic willows were the largest plants, and trees were absent. The rate of climate change was far faster than any ecosystem, save perhaps aquatic communities in shallow lakes, could respond, and for several thousand years into the Holocene period there were successive adjustments as woodland plants and animals colonised the country.

GATEWAYS AND BARRIERS TO TREE COLONISATION AND MIGRATION

Scotland has always presented a series of barriers to plants, animals and human communities. Some of these are topographical, from the upland barrier of the Southern Uplands, punctured by routeways along the east coast and through valleys like Annandale-Clydesdale and Tweeddale, the low-lying estuaries and former marshes of the central belt, and the rearing southern fronts of the Highlands themselves. The ways round the Highlands run along the seaboards. Some islands on the west coast are probably so close to the mainland that the sea formed no significant barrier, like Arran, Mull or Skye, but the Outer Hebrides and the Northern Isles may have been more isolated from mainland seed-sources.

country, around 11,000 years ago, juniper and other tall shrubs probably dominated many regions, growing luxuriantly in the absence of competing tall plants (see Plate 1.1). They experienced no climatic or soil constraints. A few hundred years after deglaciation, birch arrived from the south and east, and its appearance rapidly fragmented the cover of juniper. Our best guess is that shrub communities were driven to higher and higher altitudes as trees out-shaded them, but in truth we simply cannot trace what happened to them. It has proved a difficult task to establish that the isolated montane shrubs that we find today originated early in the interglacial, however attractive this idea is.

Birch colonised everywhere by 11,000 years ago, with no constraint on migration, save that of the distance seeds can travel. Similarly, some 800 years later, hazel also found it surprisingly easy to grow everywhere: its seeds can spread remarkably effortlessly across sea-water, so colonisation of the Outer Isles is not unexpected. Some researchers have suggested that hazel colonised mainland Scotland from areas off the present coast that had been dry land during deglaciation (see Fig. 1.2). Hazel readily developed an unstable, competitive relationship with birch. Over the next 1,500 years these early colonisers, who sacrifice the ability to compete with other trees for an ability to establish rapidly on new ground, were joined by oak and by elm from the south, slower spreading than birch and hazel but competitively superior (see Plate 1.2).

In the north of the country, the region surrounding Loch Maree was the focus for

PLATE 1.2 *A mixed oak–hazel–birch–alder woodland in Argyll. Although this woodland type has probably been altered by human activity since the Neolithic period, this mixture of species may be similar in some respects to the original woodlands. (Forestry Commission.)*

PLATE 1.3 *Within eastern Glen Affric are complex stands of Scots pine, birch and heather moor. These are considered among the finest examples of woodlands that are close to those established 8,500 years ago, although it is not clear how different they are. (Copyright: Richard Tipping.)*

the extraordinary expansion of Scots pine, from around 8,500 years ago (see Plate 1.3). Where seeds came from to establish here is totally unclear – Ireland may be our best clue – but the absence at this time of aggressive competitors like oak and elm is one important reason why pine then spread to occupy a dominant position in large parts of the Highlands. However, there must have been different seed-sources, and subsequent periods of fragmentation and isolation of different populations, because there are significant genetic differences between pines in different parts of north, east and west Scotland.

LARGE-SCALE PATTERNING AND SMALL-SCALE VARIABILITY: THE PATCHINESS OF NATURAL WOODS

By 8,000 years ago nearly all our major trees had arrived. However, in a hostile landscape and with periods of climatic instability, we need to distinguish between colonisation, establishment and expansion. Many species may have colonised locally but then lost ground as conditions changed. Alder is one species which colonised around 8,000 years ago in many localities, but which could not successfully establish because it could not compete with existing trees. Environments had to change to provide slight advantages to alder. These changes may have been natural through climate shifts or might have involved human activity in suppressing the growth of other trees. At some sites alder populations did not expand until well after 6,000 years ago. So it looks as if, in northern Scotland at least, there was a continuous 'shuffling of the deck' in terms of species composition and woodland structure in the first 5,000 years of the interglacial, an extraordinary dynamism perhaps different to the stability of woods seen in climatically less stressed southern Britain.

We can reconstruct the large-scale patterning of woodlands in Scotland at particular times from radiocarbon-dated pollen analyses. Figure 1.3 is one such reconstruction, which depicts in very simplistic terms the broad distribution of woodland types 6,000 years ago. It is difficult to differentiate what were woodlands from scattered trees, particularly at the edges of the woods in northern or upland Scotland, but, at its greatest extent, there would have been very few places from which no trees could be seen.

Figure 1.3 represents a 'snapshot' in time when woodlands were probably most diverse and extensive, but we would be better thinking of woodland development as a film where each frame changes and movement is constant. Five hundred years before or after 6,000 years ago, the map would change markedly. As an example, we can look at the fragmentation of the huge block of pine woodland that characterised much of northern and eastern Scotland 6,000 years ago (see Fig. 1.3). This forest had exceptional mobility and fragility because pine is very sensitive to climate change. Pine populations expanded north and west from 'core areas' in and near the Great Glen in the early Holocene, but had already shrunk significantly from high ground on Rannoch Moor before 6,000 years ago. The Galloway Hills supported a local pine population between 7,500 and 6,500 years ago, but this also collapsed, never to return. In the north, pine spread again, particularly after 5,000 years ago but for an astonishingly

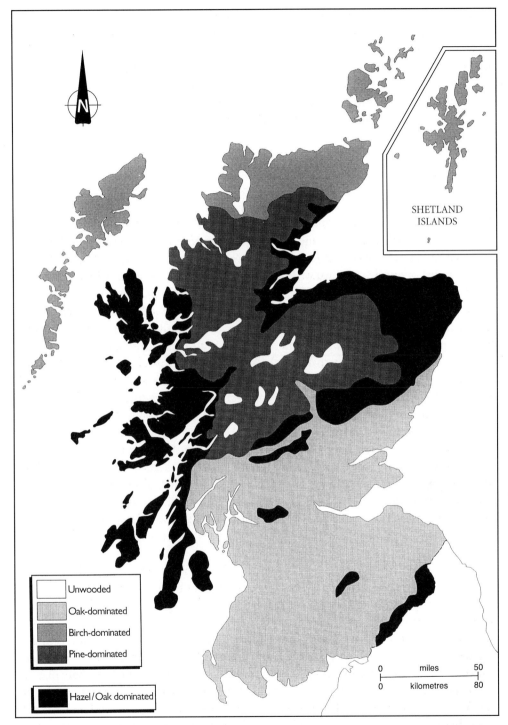

FIGURE 1.3 *A map of the major woodland types in Scotland at 6,000 years ago. (Modified from R. Tipping (1994),* Proceedings of the Society of Antiquaries of Scotland, *14, pp. 1–55 with permission.)*

short period, perhaps only for a single generation of trees, when pines reached the coast of Durness. Its spread was probably aided by blanket peat surfaces drying out because of an abrupt arid phase, but then over a comparably brief time-span this generation of trees was killed off, probably as the peats became waterlogged as the amount or intensity of rain sharply increased, and pine was irretrievably lost from large areas of northernmost Scotland.

Figure 1.3 has other limitations. We have ignored wetland species like alder or willow because they were everywhere, along valley-sides and on fens around lochs and now infilled lochans. The map depicts only those few dominant taxa within any region. We have not considered the many trees that accompanied the dominants, like rowan, ash or holly. This is partly because we know less about the prehistory of these trees. All grew within these early woods, but in more than one sense they hid behind the major taxa, quite literally in that they probably formed scattered trees and small copses and not extensive woodland stands, and in another sense they are rendered partly invisible to our palaeo-ecological techniques because they produce less pollen than other trees. Only recently, for example, have we been able to glimpse past landscapes where rowan was a co-dominant species with birch, in the Highlands, but as long as 10,000 years ago. When hazel colonised, rowan was supplanted and adopted a more unassuming role. Ash is not tolerant of shade, and would have found it hard to compete for space in early post-glacial woods, seizing advantage only on favoured calcium-rich soils. Ash has certainly gained much from the human interference that in the last 5,000 years has led to more open woods. Lime may not have been able to tolerate the cold of Scotland, although its presence on nutrient-rich soils in the south-east is suspected. Other trees certainly did not form part of the early post-glacial woodland in Scotland: beech and sycamore were introduced, probably by human agency, much later in the interglacial. We have also not discussed the shrubs that grew beneath the shade of the big trees because, again, we know much less about them. A false impression of uniformity can thus emerge from maps like Figure 1.3. Small-scale climatic, geological and topographic contrasts over short distances would have introduced an astonishing beauty, richness and diversity of woods within individual valleys.

Hunter-gatherers and Woodlands in Scotland

So far we have discussed how our woods evolved after the last stages of the 'Ice Age' without considering people at all. This is because people had very little effect on the overall course of woodland development. There have been suggestions that people may have purposefully modified vegetation at regional scales. It has been argued in particular that hazel, which was a dominant tree throughout southern and western Scotland, was encouraged by human manipulation, or even was deliberately planted in the earliest Holocene. Hazel was undoubtedly an important foodstuff to hunter-gatherers. One recently excavated shallow pit on Colonsay produced what is estimated to be a store of 30–40,000 nuts (see Fig. 1.4). We have, however, no evidence to suggest the tree was promoted by human activity, simply that it was a preferred food source.

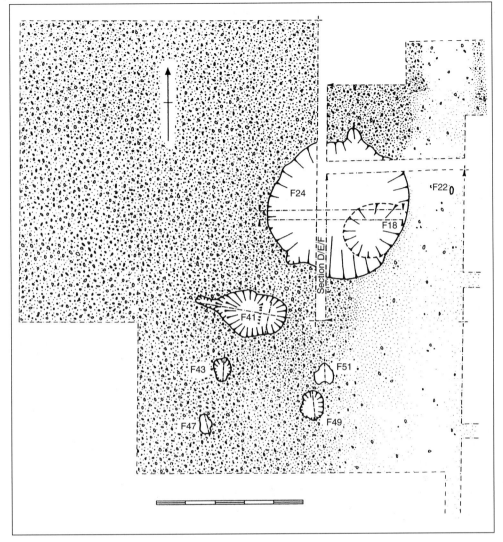

FIGURE 1.4 *A large-scale plan of the archaeological features excavated at Staosnaig on Colonsay in the Inner Hebrides. Most are small hollows, but F24 in the north-west is a 4m diameter pit crammed with the residues of hazel nutshells, thought to have been a pit for roasting this food. (From S. Mithen* et al. *(2001),* Journal of Archaeological Science, *28, pp. 223–34 with permission.)*

Later in the interglacial, poorly competitive trees like alder may have been assisted in expansion by human activities, but probably unconsciously and without intent.

People probably found Scotland at the end of the Ice Age too hostile to survive, although groups occupied caves in northern England. People moved into the country with the first trees, and this co-existence may have been more than coincidental. We have mentioned a delay between initial climatic amelioration and the earliest birch woodlands, and, as far as we can tell, human migration was similarly delayed. The

PLATE 1.4 *Archaeologists sampling organic muds for palaeo-ecological analysis at the earliest known settlement site in Scotland, at Kinloch, overlooking Loch Scresort on Rum. Drain pipes are recent intrusions into the very shallow and superficial deposits, but the density of stones in the foreground suggests that the latter were moved by Mesolithic communities to prepare a drier or more stable floor. (Copyright: Richard Tipping.)*

earliest occupation seems to have occurred around 9,500 years ago, on the east coast of Rum in the Inner Hebrides (see Plate 1.4). If this truly represents the earliest settlement, it is noticeably younger than might be anticipated on purely climatic considerations. Although we don't know where these communities originated, it may be that people followed animals that moved north and west with the migration of birch and hazel: more tentatively, perhaps people followed hazel.

The presence of hunter-gatherers is most commonly established from their stone toolkits, and this period is called the 'middle Stone Age', or Mesolithic. Highly skilled techniques were developed for shaping tools from beach flint, chert, rare fine-grained volcanic rocks from which tools were nonetheless widely 'traded', and more intractable quartz. In other parts of north-west Europe stone-working techniques were altered very early in the Holocene and characteristic tools called microliths began to be made. This change in technology probably represented a change in hunting techniques as woodland cover increased. Microliths may have been used as barbs on arrows to be fired at animals over reasonably short distances, when the hunter was partly concealed by trees. Since nearly all Scottish hunter-gatherer sites are dominated by microliths, it is very likely that our earliest communities were woodland users, unfamiliar with hunting on open grassy plains.

Mesolithic sites are very difficult to find because most are simply open-air stone tool scatters. More distinctive but more rare are piles of shellfish, animal and bird bone and 'domestic' waste called shell middens, which are exclusively coastal, while cave or rock-shelter sites are also rare and also found on the coast. The distribution of these sites was in the past interpreted as showing that occupation was exclusively coastal, partly because of the relative abundance and diversity of resources here, and partly because the wooded interior was pictured as almost impenetrable. Now we recognise this distribution is skewed by many factors, and findspots have been found far inland, high in the Galloway Hills and in upper Tweeddale, for example. Woodlands were clearly not entirely impenetrable, a point returned to below. It has proved very tempting for archaeologists to link together different types of site, between coasts and inland, for example, or between what are seen as base-camps and upland hunting settings, and complex models have been developed which stress seasonal movement and mobility of groups, but these models are far from proven.

It is often difficult to understand what was being hunted, but at the rare sites where animal bone is preserved, such as at Morton, near St Andrews on the east coast, the game were woodland fauna, like red deer, which were much larger than the animals we see today, roe deer, boar and aurochs (large wild cattle). We need not think that hunting was entirely opportunistic. These communities, of course, planned their hunting strategies. Some researchers have in recent years suggested that the management of wild resources went well beyond taking advantage of chance encounters by suggesting that red deer were purposefully introduced by boat to some islands such as Shetland, the Outer Hebrides and the Inner Hebrides. Islands like Colonsay in the Inner Hebrides have good collections of red-deer bones in shell middens from this period, but deer could have swum to these unaided and simply been hunted there. It is much harder to envisage deer swimming to Shetland or Lewis, and if the presence of deer in the Mesolithic could be demonstrated on these islands, it would imply that hunter-gatherers were capturing and corralling live animals in ways that are very close to farming. No bones of large mammals are known for this period from these islands, however, and interpretations have instead focused on inferred impacts of grazing animals on natural woodland vegetation as seen in the pollen record. This is a controversial issue because we have to infer much from quite scanty lines of evidence, but the idea is exciting in emphasising that these people were far from primitive.

Further sophistication in how Mesolithic communities might have used their woodland landscapes, though again controversial, also comes from interpretations of the pollen record. At some localities, often in upland settings within oak–hazel woods, we can identify short-lived events, each lasting only for a few years, in which small grassy patches in the local woodland were created. Fire was repeatedly associated with these reductions in the extent of woodland because charcoal is recorded in the sediments, and what replaces trees temporarily was grassland. These events can happen several times at one pollen site, and the best work, from south of the border in the North York Moors, shows very elegantly how open grassy patches within woodland across one hillside were created and then healed by woodland regeneration at different times. Human activity has been postulated to have caused these modest but repeated disturbances. The preferred model is that woodland clearance by fire leads to the encouragement of grasses, and this newly created grazing attracts wild grazers like deer or aurochs. Not only are these animals fattened on this fresh grazing, but they will return consistently and predictably to these patches, making them easier to hunt. This is an exciting interpretation, and has some support from observing present-day hunter-gatherers, but we are stretching our levels of inference to the full in this reconstruction, particularly when other mechanisms can account for the same ecological patterns. The explanation of woodland loss through purposeful human-set fire is neat, but only if we assume that fires cannot be generated naturally, through lightning strikes, for example. If they can, human involvement is an unnecessarily complex explanation. So the current debate centres around whether these woodlands can have been set alight naturally, and we simply do not know this. Much depends on what the woodland composition was, which was highly variable in space and time as we have seen, as well as on the soil, especially whether organic matter can accumulate on forest floors, and whether the prevailing climate allowed organic matter to become dry enough to serve as tinder. The need for such interference may have been more in the oak–hazel woods of southern Scotland and yet entirely unnecessary in the open birch woods of the far north.

Most of the effects that researchers have inferred for hunter-gatherer impacts on woodlands are difficult to test because they are really only extensions of natural disturbance processes. So anthropogenic manipulation of woodlands, if it occurred, only replicated the sort of patch-creation that people observed after, for instance, a severe storm had knocked down trees, or after a forest fire, or during the creation of large open patches by the activities of beaver (see Plate 1.5). People in the Mesolithic seem to have had little requirement to do more than mimic nature. Even campsites made almost no impact on their surroundings. Stone buildings are not found: stones were

PLATE 1.5 *A small beaver dam in Patagonia, South America, which has created the pond behind the dam, drowning the trees and creating an open patch in the woodland. (Copyright: Richard Tipping.)*

only moved to form small circular settings. Because of their mobility, hunter-gatherers probably most frequently used lightweight tents of skin supported by cut branches. Their obsession with hazel has been alluded to already, but other woodland products like acorns, birchwood and bog myrtle (not correctly a woodland shrub), and the appropriation of fruits as with pear at one site in Northern Ireland, rendered no permanent damage to the wood. After decades the woods closed over, although disturbance would probably have altered temporarily or permanently the mix of plant communities. Light-demanding trees like hazel, birch and ash, and shrubby taxa like alder and thorns, may have grown more profusely in modified woods, and would have appeared different to the people moving through them.

Walking in and around the Woods

What did being inside the woods feel like? This may have seemed a confusing question to our hunter-gatherer ancestors because it implies that there is an outside to the wood, from which you have to enter. This is how we perceive woods these days, as isolated habitats that have boundaries. Until late in prehistory over large parts of Scotland, people were probably always inside the woodland, except on coasts, along rivers (though here the floodplains would support wet woodland) and on montane plateaux. This was where they lived, within and belonging to woods.

We are not very good at measuring how it was to walk through these woods, but it is clearly an important issue. Reconstructions of dense forbidding forests, impenetrable to people, led to ideas of Mesolithic communities keeping to the coasts as 'strand-loopers'. Some recently recovered data from sites where the original woodland structure is well preserved, from landscapes where spreads of sediment or peat have buried trees, suggest a similar image, with tangled brambles and thorns blocking the way, and trees growing very close together, cutting out the light to the ground-flora. However, the conditions required for preservation mean that these examples may not be typical of most of the original woodland. The same problem occurs with what appear to be our best examples of preserved forests, the extensive remains of pine stumps within now-eroding peat throughout northern and western Scotland (see Plate 1.6). These are 6,000–4,000-year-old forests, but careful work using tree-ring analysis has shown that individual trees found next to each other are not of the same age. These trees were not all growing at the same time, so that the density of stumps visible in the peat is higher than the numbers of trees that were in the original woodlands.

Walkers today probably find the tangle of thorns and brambles most depressing obstacles, and these were always present. In season, they would have provided fruits

Plate 1.6 *The remains of 6,500-year-old pine stumps emerge from eroding blanket peat at Clashgour on the south slopes of Rannoch Moor. They give the appearance of being a beautifully preserved forest, but tree-ring analyses have shown that few if any trees grew side by side. The pine woodlands to which these trees belonged was lost through climate deterioration, 'drowned' by the waterlogging of the peat that rapidly buried and preserved the trees. (Copyright: Richard Tipping.)*

Plate 1.7 *Open oakwoods on Loch Lomondside. These woods are regarded as semi-natural, but their composition and structure have been almost entirely restructured by human activity in the last few hundred years. (Forestry Commission.)*

and berries, of course, but out of season they would have been decidedly off-putting. We associate these shrubs with woodland edges, but often today these shrubs have been planted in hedges to act as barriers to animals as well as walkers, to stop us entering the wood. We thus gain a false impression that walking within woods might have been relatively unimpaired. Many of our present-day woods are even-aged because they grew after single massive disturbance events such as human clearance, but natural woodlands were mixed-age, with small patches of trees called stands living and dying together. Death and decay would allow the sideways movement of neighbouring stands, not necessarily of the species that had died, as light encouraged new seedlings. Edges inside woods will have been common, and small gaps abundant because the different-aged patches of trees would die and decay at different times, and the same tangle of adventitious thorns and climbers may have restricted movement more than we think.

The image of woodland interiors looking like, for instance, the New Forest or Loch Lomondside today (see Plate 1.7) is probably incorrect, then. These woods are open and verdant because of unnaturally high grazing pressures or because of former land uses: they are artificial. Some ecologists have sought to draw parallels between our native woodlands and the open deciduous woods of forest reserves in Europe or the grassy glades seen by the Pilgrim Fathers in eastern North America, but often we have underestimated the amount of anthropogenic manipulation in environments we fondly but mistakenly think of as 'natural'. Equally, ecologists have suggested that wild grazing animals would have naturally created the open grassy conditions of reserves like the New Forest, but these wild animals had more predators than just humans, and it is doubtful that natural grazing pressures ever exceeded the capacity of the wood to regenerate. Original woods were probably much more cluttered than the image in Plate 1.7. Dead wood would have dominated the floor as trees died and fell: our woods today are no analogue for these because most are tidy and logs are removed. Edible mushrooms, on the other hand, will have been more abundant than today.

Although we have comparatively abundant evidence for Mesolithic communities from inland locations now, many sites are situated at the edges of woods, either on the coast, overlooking or on rivers, or in the uplands where trees may have thinned. Manipulation by fire is one way of creating edges within woodland. These edges or ecotones were important because they represent borders between different sets of resources. Mesolithic people understood this, and their need to go deep into the woods may be overestimated. It is likely that long-distance travel was by boat, along rivers and coasts, rather than on foot. So as well as small patches in woods, we can think about whether there was an outer edge to these woods, and what was outside the woods. The more extreme climates of northern Scotland have meant that this has been seen as the best place to find a northerly edge, but in the last few years researchers have found that trees grew, at least for a time, almost everywhere. The pollen representing local trees has been found as far north as Shetland and as far west as the Uists: probably even St Kilda had birch trees in the middle of the Holocene 5,000–4,000 years ago. It is more difficult to know whether these trees could be called woodland.

Researchers have also sought an upper edge to the woods. There is today only one location high in the western Cairngorms at Creag Fhiaclach (650m above sea level) that can be thought of as a natural tree-line. In the past, the Southern Uplands may have been entirely tree-clad in the later Mesolithic, around 7,000–6,000 years ago. In the Moffat Hills, at a site delightfully called Rotten Bottom, 600m above sea level, the oldest bow yet found in Britain was pulled from peat. The flatbow, made of yew probably from trees growing in the northern Lake District, is radiocarbon dated to 6,000 years ago. It was initially thought to represent a hunting trip into the uplands above the woods (a failed one since the bow snapped and was probably thrown away in disgust) (see Fig. 1.5), but analyses have found that birch, hazel and oak trees probably grew at this altitude then, and hunters may have needed this light woodland cover to get close to their quarry. Further north, estimates for tree-lines of around 700m above sea level have been made for the Eastern Grampians and 500m above sea level for the north-west Highlands. These estimates are probably maximal and with changing climates tree-lines would have been suppressed: so the upper edge to woods expanded and contracted over time.

A NEW WORLD: THE INTRODUCTION OF FARMING

Some 6,000 years ago, the world changed for hunter-gatherers in Scotland. The means of production of food altered because domesticated livestock and crops were made

Figure 1.5 *A reconstruction by Marion O'Neil of a 6,000-year-old hunting trip to Rotten Bottom in the Moffat Hills, made soon after discovery of the oldest bow in Britain, seen here in the foreground. Detailed reconstruction of the landscape has suggested that rather more trees than are depicted here might have provided good cover for the hunter. (Reproduced by courtesy of the Trustees of the National Museums of Scotland.)*

available, having been transferred across Europe from the Near East, where agriculture had been invented at the end of the Ice Age. Neighbouring communities on the continent had farmed for perhaps 1,000 years before it was introduced to the British Isles. This suggests that the transition to an agricultural world, the adoption of a Neolithic ('new Stone Age') economy, was not automatic, and that there was resistance to new ideas within Mesolithic communities. In southern Scandinavia it seems that hunter-gatherers were similarly choosy, selecting elements of agriculture but not necessarily buying into the entire package. Their existing lifestyle was familiar and seemingly successful. We have had little evidence for crises developing which forced the adoption of agriculture, but very exciting new data from the Inner Hebrides suggest that suddenly 6,000 years ago Mesolithic settlers abandoned their predominantly marine diet. Something dramatic in the climate may have happened, and we are beginning to look to changes in the North Atlantic Ocean currents for explanation.

Certainly very soon after 6,000 years ago, substantial funerary monuments were being constructed in stone, chambered tombs found in the north and west of Scotland. Conventionally such monuments were seen as being constructed when an agricultural economy had become proficient enough to provide a food surplus and time to think, but these sophisticated buildings date to the earliest Neolithic period, too early to represent the signature of fully-developed economic achievements. Instead, they are increasingly seen as symbols used by people beginning to explore control of their environment, with a new-found confidence in themselves and the potential for mastery over the landscape. Thus, building in stone differentiated early farming communities from their hunter-gatherer ancestors, who did not think in terms of altering the earth, even though they may have been essentially the same communities. Mesolithic people seem to have wanted to move quietly over the earth without disturbing it: Neolithic communities could change things, and sought to show this.

Early farming communities had the power to change landscapes because they had new toolkits. They had large stone axes, and elegant experiments in the 1950s showed that such hafted axes were very effective at cutting down trees. Hunter-gatherers did not have this technology, and perhaps did not have the desire to cut trees in this way. Again we need to think of stone axes in a symbolic sense. Many axes were manufactured from attractive stone, exhibit very high craftsmanship, are polished and clearly were not actually used. They too were statements. We have no clear evidence from the pollen record of mainland Scotland that extensive clearings were created in woodland in the Neolithic period. The axe was added to fire as a potential way of clearing ground; indeed, there is some evidence from south-west Scotland that fire was used less now. But we have no extensive openings in woodland except on the Northern Isles. On Orkney and Shetland the woodland was substantially lost in the Neolithic period, although it may not have disappeared, but it is unlikely that this was entirely due to human impacts. Woodland this far north was probably highly stressed, and even small impacts may have been enhanced by the already tenuous hold of trees on this landscape. What appears to have changed throughout the British Isles, however, was that gaps in the woodland, though no larger than in the Mesolithic, were more permanent. Clearings may have remained open for hundreds of years. This is a significant contrast because woodlands were not allowed to regenerate. Seedling establishment was probably prevented by directing livestock like the newly-introduced sheep, or pigs, to fresh shoots as much as by ploughing seedlings into the ground.

Simple stick-like ploughs called ards were used to break turf for the first time. This too had its symbolism. New crops like strains of wheat and barley were available, but there is heated debate at the moment over whether crop-growing was taken seriously in the early–mid Neolithic, before about 5,000 years ago. Although it makes sense to interpret as part of the food economy the charred remains of cereals on archaeological sites or the pollen record for crops, one current view is that early farmers did not rely on cereals and regarded them largely as potent symbols of what they could do.

This debate centres on the level of 'investment' in agriculture of these farmers and whether they were entirely committed to farming. In Shetland, perhaps Orkney, and in western Ireland there were organised cohesive field systems with stone walls, and it is easiest to see these as representing a serious commitment to farming. Elsewhere, especially in southern Britain, the evidence is less pronounced and this has led to the suggestion that early farmers remained nomadic, not wanting to give up their old hunting ways: hazelnuts, for example, continued to be collected from wild places.

This latter view has been encouraged by our difficulties in finding the settlements of these early farmers. In large parts of Britain there is little archaeological evidence for permanently occupied houses despite the observation that crops, for example, should need tending. On Orkney and on North Uist are small rectangular huts – that at Knap of Howar on Papa Westray has a timber post from a species of conifer – but whether permanent, we do not know. One so far unique building in Scotland is at Balbridie in the Dee Valley, where a large (26m x 13m) early Neolithic timber hall was excavated. The nearest parallels to this building are on the Continent. Frustratingly, although wood fragments were ubiquitous, it is not clear what timber this large hall was built from.

NEOLITHIC AND BRONZE AGE WOODLANDS

The extent of woodland in Scotland probably did not diminish significantly between 6,000 and 5,000 years ago. Significant changes occurred in the species composition of deciduous woodlands south of the Great Glen, and probably further north in sheltered areas, but it is not clear whether these have an anthropogenic cause. Most dramatic is the loss of elm trees from the former oak–elm–hazel woods. This event affected every part of the British Isles where elm grew, within a few hundred years of 6,000 years ago, when populations of elm trees died. In some places elm grew back a few hundred years or so later, though not abundantly, and in others elm was entirely lost. This was until recently seen as the earliest event in the British Isles that a tree species had been lost – we too have emphasised the processes of migration and species recruitment before this time – but in Scotland earlier declines in pine populations are known (as mentioned above). Viewed in this perspective, the decline in elm is less special. Climate change is a plausible mechanism for the losses of elm, but there are others. Scientists can be creatures of fashion, and in the 1970s Dutch Elm Disease was seen as the model for a similar disease 6,000 years ago. Human involvement in early agriculture has been seen as likely, because elm is known to be eaten selectively by grazing animals and domesticated livestock would need to be fed over winter, always the critical period in a pastoral economy. Elm was thought to have been heavily predated by people lopping branches to feed to sheep or goats until the resource was exhausted, but such mismanagement of a critical resource is unlikely. In addition, the simultaneous cropping of elm throughout the country implies a prodigiously large early Neolithic population, for which there is no evidence. Nevertheless, human agency is not easily dismissed for specific sites: on the coast at Oban, wood charcoal from a cave site occupied at the transition from Mesolithic to Neolithic economies included 40 per cent elm wood, which probably represents at least its selective collection, even if we do not know the reasons for selection. We do not know what caused the elm decline, after sixty years of patient enquiry. There remain several competing hypotheses, and many workers now prefer an explanation where several factors combine.

Around 5,000 years ago the use of agriculture gradually became more discernible in the landscape. Sustained but not necessarily intensive grazing pressures probably

PLATE 1.8 *An aerial photograph of the heather- and peat-covered moorland south of Lairg in Sutherland, showing in the foreground the excavated areas revealing the round outlines of Bronze Age hut-circles. (Copyright: Rod McCullagh.)*

achieved more work than the axe or fire in creating partly cleared landscapes. It is very unlikely that significant parts of the country were converted to permanently altered open ground by human impacts, however. It is more likely that climate change and soil deterioration were more significant agents, particularly in northern Scotland. Here the catastrophic collapse in pine populations 4,000 years ago, described earlier, opened up entire regions in Sutherland and Caithness which have remained virtually treeless since then. The conversion of mineral soils to nutrient-deficient acid blanket peat had occurred here and elsewhere from the beginning of the interglacial, but rapidly increasing precipitation waterlogged this natural sponge, forcing trees from its surface through cutting off air to roots (see Plate 1.6). As the amount of 'dry' ground dwindled, woodland shrank to increasingly isolated patches on 'islands' surrounded by peat, which may have become so stranded that the supply of seed was reduced below viability.

This extraordinary loss of trees occurred when parts of the Highlands were receiving influxes of settlers, in the early Bronze Age. Throughout the straths and hills of northern Scotland are the turf and stone remains of their hut-circle houses, accompanied by small plots of tilled ground among the rough pasture and heath (see Plate 1.8). The idea that this wave of colonisation might represent a large population expansion in some 'golden age' of climatic warmth has been replaced by the opposite notion. Climate-change history shows this period to have experienced the most severe deterioration of any in the last 5,000–6,000 years. What were people doing in the hills at that inhospitable time? New interpretations stress the close inter-relation between

PLATE 1.9 *The exceptionally well preserved stone village of Skara Brae on the west coast of Mainland Orkney. Wood charcoal recovered from excavations has shown that scavenging for driftwood on the nearby beach provided a large proportion of building timber and fuel. (Copyright: Richard Tipping.)*

PLATE 1.10 *A photograph of a driftwood-strewn beach in the Magellan Straits, Patagonia. The abundance of wood comes from the extensive and uncut southern beechwoods around this bay. (Copyright: Richard Tipping.)*

woodland decline and the potential for agricultural success. In west Glen Affric, east of Inverness and over 250m above sea level, woodland collapse appears to have commenced just prior to the first human settlers, and this may have encouraged farmers to utilise the expanded grazing land despite deteriorating climate. Once there, farmers had sufficient robustness and ingenuity to remain, although faced with equally harsh conditions in later prehistory.

Against the large-scale images of regional woodland collapse over large parts of the northernmost mainland, we can see small vignettes generated from the detail of individual archaeological sites. From the west mainland of Orkney at the 'classic' stone village of Skara Brae (Plate 1.9) we find that by the late Neolithic, around 4,400 years ago, over a quarter of all wood charcoal came from driftwood, many from east North American coasts. This careful observation emphasises the paucity and impoverishment of the Orcadian woodland that we saw commenced 1,600 years before. Further, it shows that the very large forests across Europe and North America would have produced huge amounts of driftwood on beaches (Plate 1.10).

At the opposite end of the country, Lintshie Gutter in upper Clydesdale is a Bronze Age cluster of platform settlements, the Borders equivalent of northern hut-circles, and these predominantly timber houses were manufactured from many species, but almost exclusively of smaller scrub taxa like hazel, alder, willow, birch and rowan. Oak was a rare component; elm was, as we can now appreciate, absent from the area. This site is intriguing because regional pollen records suggest that although the proportion of scrub elements was increasing gradually in later prehistory, the woodland was close in species composition to pre-Neolithic woods. Clearly there was either selection of wood at Lintshie Gutter, and the puzzling avoidance of that excellent building timber, oak, or we have much to learn in the comparison of different sorts of information on our past woodlands.

INTO THE IRON AGE

This chapter finishes not at the end of prehistory and with the arrival of Roman troops, but a thousand or so years before this at the beginning of the Iron Age. This subdivision of the prehistoric record is intentional, because in some areas of Scotland the Iron Age sees a quickening of the pace of change in the archaeological record. Some aspects of the relation between people and woods become better resolved, and in southern parts of the country there developed an approach to landscape-scale deforestation in farming communities that had no parallel in earlier times. Prior to the middle of the Iron Age, around 2,500 years ago, the involvement of people with their woods had been an intricate one, a symbiosis rather than the confrontation that we often think. Even the imposition of human will over nature that the Neolithic period has come to represent did not in any significant way distort our relation to woodland and trees. Both trees and people shared the same vulnerability to the greater driving force of climate and landscape change. Perhaps it was this, combined with the knowledge that since the beginning of time each had shared the other's landscape, that allowed enduring respect. In large measure, this intimacy was lost in the coming hundreds of years.

CHAPTER TWO

The Coming of Iron, 1000 BC to AD 500

Ian Armit and Ian Ralston

Introduction

The beginning of Scotland's Iron Age is hard for archaeologists to discern, as many of its conspicuous elements – notably settlements of roundhouses – strongly resembled types in use in the preceding later Bronze Age. The evidence for iron itself is muted. Fragments occur in a few collections of late Bronze Age metalwork, and the earliest iron-working seems very limited in scale. The increasing use of this readily-available ore is suggested because copper alloy cutting tools went out of production. The Iron Age began in the eighth century BC; it continued into the first millennium AD, across the centuries when Roman military forces were present.

In the final centuries BC, iron production probably increased in importance. The impact of iron tools on Scotland's environment, especially on woodlands in the south, can more readily be discerned. During the 'Roman period' – in reality a series of brief interludes of military intervention – a large foreign army was intermittently stationed across the southern half of Scotland. In some views, the specific requirements of this force are suggested to have altered Scotland's environment considerably. Changes during this span – from the later first century AD until the third, and (more tenuously) to the beginning of the fifth – may relate more to economic and political influences from the Roman Empire to the south than to the presence of garrisons here.

During the Roman period, the population continued to consist largely of indigenous stock and it was their elites, most probably Celtic-speaking, who emerge as the political leaders at its end. The post-Roman Britons, Scots and Picts represent in large measure a continuation from the pre-Roman Iron Age. Later in the first millennium AD, the arrival of the Angles, and subsequently the Norse, introduced Germanic speakers and increased the likelihood of internal disruption within Scotland. The rising importance of Christianity contributed to radically changed perceptions. The survival of a contemporary documentary record is an important influence on the nature of reconstructions. For these reasons, the sixth century AD marks a convenient end-point for this chapter.

The Nature and Extent of Woodland Cover

By the beginning of the Iron Age (c. 800 BC), the inhabitants of Scotland were living in an environment already much altered since the country's initial colonisation. These changes arose as a result of human acts, deliberate and involuntary, and climate change. As a general rule, the influence of climate seems to have been more marked in the north and west, where higher rainfall, increased exposure to wind and locally to salt spray, and similar unfavourable conditions deriving from the influence of the Atlantic Ocean contributed markedly to environmental deterioration.

A key change due to both anthropogenic and natural processes was the substantial reduction of forest cover. It is, however, presently impossible to give a national picture of the extent of woodland cover at any date within our time-span. Nonetheless, it is clear that the decline from an estimated 50 per cent natural woodland cover in the whole of Scotland at the beginning of farming in the fourth millennium BC to some 4 per cent by the eighteenth century was already well advanced.

Evidence from pollen analyses remains very important, as it had been for earlier prehistory, but the inferences drawn are not always incontrovertible, dependent as they are on the resolution of the analyses and the estimation of a timescale from, in some instances, limited numbers of radiocarbon dates. Palynologists also differ in their interpretations as to what constitutes 'complete' or indeed 'extensive' clearance. The apparent decline of woodland may imply pollarding and coppicing, processes that reduce flowering and hence pollen production, rather than the wholesale removal of trees. Coppicing, leading to the production in a few years of relatively straight-growing poles useful for building, is most likely to have affected hazel and ash; it can, however, be practised on other tree species. Although each of these processes impacts on pollen production, the broad pattern of Scotland's forest cover during this span can still be sketched. In what follows, it should be noted that even clearance would have often entailed maintaining an appropriate local woodland resource.

Another source, much used in previous histories of Scottish forests, are the descriptions in the first written accounts on the country, composed by Roman authors, but these brief statements are problematic. Caledonia's reputation as a heavily wooded country in classical sources composed from the first century AD onwards may have been exaggerated in the minds of southern authors familiar with very different landscapes. Perhaps these were also in part a literary device to highlight, alongside mentions of marshes and bogs, the strangeness and difficulties of the northern terrain, and the soldiering skills of Rome's commanders.

In general, the species composition of Scotland's woodlands during this period, as recovered primarily through pollen studies, remained unchanged from earlier prehistory. Such exotic species as are represented may be explained as driftwood originating on the seaboard of North America, or by the long-distance dispersal of pollen from trees growing elsewhere. While, in southern Britain, the Roman period coincides with the introduction or spread of species of tree previously either relatively rare or absent in the country (such as hornbeam, sweet chestnut and some fruit trees including plum), there is no evidence to suggest that any of these were successfully established in Scotland at this time. Tools associated elsewhere with the tending of orchards do, however, make a rare appearance on Roman sites here, as is discussed below.

The Western and Northern Isles seem to have lost the vestiges of such woodland

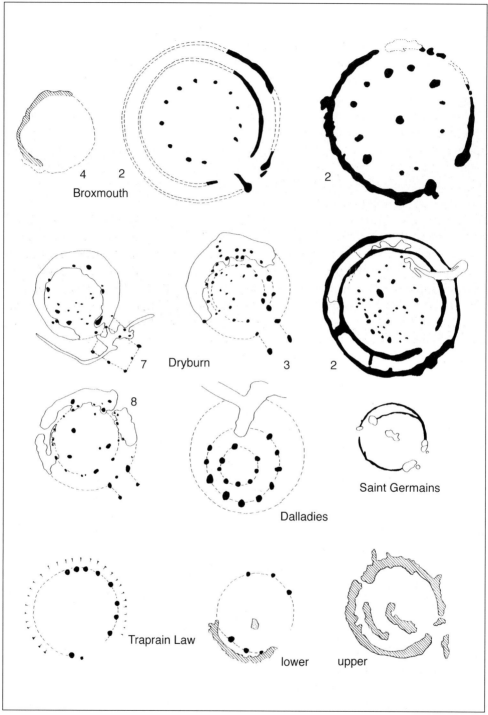

FIGURE 2.1 *The plans of the earthfast elements of timber roundhouses are frequently recovered on Scottish Iron Age settlements. The largest of these eastern Scottish examples is about 18m in diameter. (Sources: various, redrawn by Gordon Thomas.)*

PLATE 2.1 *The full-scale reconstruction of a timber roundhouse at Archaeolink in Aberdeenshire gives an impression of their original appearance. (Courtesy of I. A. G. Shepherd.)*

cover as they once possessed by the start of the last millennium BC. In contrast, on the northern mainland, the decline does not seem to have gathered pace until in some instances (such as eastern Caithness) the middle of the first millennium BC. Elsewhere in the north, the gradual disappearance of substantial tracts of what was probably relatively open woodland may not have been underway until the first millennium AD, or, in some cases, more recently. In certain areas, notably in Wester Ross, the palyno-logical record suggests that human impact on forest cover during our period was minimal, but this pattern is not replicated throughout the west coast of the mainland. Around Loch Shiel, for example, there may have been substantial clearance (marked also by significant soil erosion) in the last few centuries BC. The Inner Hebrides appear to have lost much of their woodland cover by the early part of our span, as at Trotternish on Skye. The pollen record from other islands indicates that considerable inroads had been made into their woodland, in some cases accompanied by signifi-cant indications of agriculture, as in eastern Mull, in the first millennium AD. But variability is to be expected on the islands as elsewhere, and, in other cases, birch scrub can be shown to have re-established itself on former pasture-land, for example on Colonsay, early in the Iron Age.

The survival of extensive forests into our period seems also to have characterised some of the valleys running into the Cairngorm massif. Here, more substantial clearances of the predominantly coniferous (Scots pine) woodland began to occur within the Bronze Age, sometimes followed by the regeneration of woodland in the first millennium AD. Elsewhere in the southern Highlands, for example around Rannoch Moor, the change from forest to peatland seems to have taken place naturally, ensuing from changes in rainfall and soils, with minimal signs of human effects.

In other sectors of the southern Highland margins, the seasonal movement of livestock has been suggested as the principal non-natural cause of slight reductions in upland woodland cover. In the north-eastern Lowlands, evidence from the Howe of Cromar, a basin on the Highland margin in Aberdeenshire, points to extensive clearance for the better part of a millennium from the start of the late Bronze Age, which was accompanied by the inwash of considerable quantities of soil into nearby lochs. Radiocarbon dates suggest that regeneration of the birch/hazel woodland here occurred late within the pre-Roman Iron Age, and the intensity of human activity in this area in the second half of our span seems to have been much reduced. Other sites showing substantial clearance within the first millennium BC include Black Loch, above Lindores in Fife, which, despite the absence of archaeological evidence for contemporary human occupation, was surrounded by a substantially open landscape during this period.

In southern Scotland, the removal of woodland was already underway, as elsewhere, from the Neolithic, but the middle of the last millennium BC appears to be a period when evidence for clearance intensifies significantly. While clearance was already underway in the Galloway Hills, for instance, before this date, a substantial upswing in the pace and extent of woodland removal in both the Southern Uplands and the Midland Valley occurs thereafter. Considerable tracts of southern Scotland – in common with northern England – were largely deforested in a series of rapid events in the immediately pre-Roman centuries, in a wholesale manner not previously encountered. Pollen evidence with supporting radiocarbon dates comes from Letham Moss, near the River Forth, for example. This trend represents the northern extension of temperate European agricultural regimes that were both intensifying and diversifying and which affected wider areas in the later pre-Roman Iron Age, from about the third century BC. Even in southern Scotland, however, some blocks of woodland will have survived.

What impact did the Roman presence have on Scotland's woodlands? In recent years, two contrasting views have been propounded. On the one hand, it has been suggested that some of the southern Scottish clearances were brought about not by Iron Age farmers but by the extraction of timber for Roman military installations; on the other, that from Fife northwards on the east, woodland regeneration episodes at various localities, broadly contemporary with the Roman period, are attributable to political and economic dislocations, and perhaps consequent depopulation, caused by the Roman military forces. Neither of these suggestions is universally accepted. Large-scale clearance, as Tipping has demonstrated (see Fig. 2.3), had got underway before the Roman incursions and may well have continued through the early first millennium AD without substantial modifications due to Roman military demands. Recent work, for example at Crag Lough near Hadrian's Wall, supports the view that the linear frontiers of Roman Britain were constructed through landscapes that were already substantially cleared of trees. In the case of the Antonine frontier between

Clyde and Forth, this perspective is supported by the substantial requirements for turf (itself necessitating land cleared of trees) that the construction of this wall would have required. Furthermore, the calculation of the timber requirements of Roman military works suggests the exploitation by widespread felling of areas perhaps only three times the extent of the enclosed sites themselves.

Regeneration (the re-establishment of woodland in a given area after clearance and subsequent land use such as for pasture) is certainly documented at some sites, such as Black Loch, Fife, broadly contemporary with the Roman incursions. Owing to the imprecision associated with radiocarbon dating, it is difficult to be sure that the reversion to woodland noted in parts of eastern Scotland was directly the outcome of Roman pressure. During the first millennium AD, woodland regeneration has been remarked in a number of localities, for example relatively close to the (by then) obsolete Antonine Wall. It has been proposed that the later military withdrawal from Hadrian's Wall prompted the reduction of agricultural land in its immediate vicinity, in turn leading to woodland regeneration in such localities as the surroundings of Burntfoothill Moss near the head of the Solway. However, other sites show indications of greater openness during the middle part of the first millennium AD; these include Black Loch in Fife, where regeneration had occurred earlier in the millennium, and the lowland Bloak Moss in Ayrshire.

IRON AGE FARMING AND THE WOODS

In considering the changes occurring in Scotland's woodlands, the activities of the human communities which played a substantial part in reducing the amount of cover must be evaluated. There is no ready way of reckoning the size of the native population at any time during these two millennia. The density of settlement (see Fig. 2.2) and the size of the greatest of the Roman armies garrisoned in or near Scotland – perhaps totalling around 30,000 men – suggest that the figure may have been of the order of a quarter million, maybe even higher (the population of Roman Britain has been suggested on similar flimsy grounds to have reached several million). This is only a guess.

The buildings of Iron Age Scotland (see Plate 2.1 and Fig. 2.1) provide the readiest archaeological indication of managed woodland, for experimental work shows them to require quantities of roughly similar-sized pieces of wood – from thin and flexible hazel rods for wattling to substantial oak timbers. Roman buildings relied more extensively on scantling, roughly squared timbers, but there is no justification for envisaging stockpiles of ready-seasoned wood, heralding the construction of a new suite of forts. The army probably obtained the vast bulk of its timber supplies locally. It is debatable whether wood with the qualities required for both the native and Roman styles of construction could have been readily and recurrently obtained from 'wildwood', and this is used as an argument for the existence of managed woodlands. Fuel for heating and as an energy source for crafts could have been exploited from natural stands, but again may have come from managed woodland.

While it may seem dangerous to make reference to Irish first millennium AD evidence in regard to Scottish woodland, these are the nearest detailed documentary sources chronologically and geographically to our period and area of interest. Irish sources suggest the treatment of wood and woodlands was very sophisticated. There was a complex gradation of trees by species and the likelihood of private ownership of much woodland (although this is much more debatable for Scotland). The economic importance of wood and other woodland products is also detailed in these seventh-century and later laws.

A driving force behind clearance would have been the requirement to obtain land for grazing and cultivation. Despite the traditional view of Scotland being dominated by stock-raising at this time, with 'Celtic cowboys' ranging freely over undivided upland pasture and moor, the available evidence suggests that most agricultural regimes involved a complex admixture of livestock farming and the cultivation of cereal and other crops. The physical remains of prehistoric field systems survive in varying condition in several Scottish landscape settings, from the present arable lands along the Esk in Midlothian, where remarkable evidence is preserved as cropmarks, to the upper slopes – at almost 500m above sea level – of Ben Griam Beg at the head of Strath Halladale in Sutherland (see Fig. 2.4). Extensive systems of low stone dykes edging small fields and sinuous trackways have been mapped on lowland Dinnet Muir in the Howe of Cromar, Aberdeenshire, and also in upland Perthshire, for example. In the Moray uplands, at Tulloch Wood south of Forres, there are signs of a planned field system laid out on two axes, and thus, though much smaller, resembling the extensive field systems covering thousands of hectares on Dartmoor. It is probably of late Bronze Age date. The northward-draining catchments from the Cheviots have revealed extensive areas of what seems to be late prehistoric terrace cultivation, perhaps preceded by more restricted high-altitude plots of spade-dug narrow rig, rarely individually greater than half a hectare. In some areas, extensive systems of linear earthworks have been noted: an early first millennium AD cattle ranch is suggested as the reason for the earthwork banks running outwards from the earlier hillfort of Castle O'er in eastern Dumfriesshire. On the other hand, many areas of Scotland, especially in the eastern Lowlands which have plentiful evidence of habitation sites, do not show clear evidence for early land and field divisions, although these may have been achieved using hedges or other devices difficult to discern by archaeological means.

Extensive as some of these systems are, however, they are universally smaller than a square kilometre in extent. What Scotland seems to lack are the really big field systems, or the lengthy 'ranch boundaries' dividing terrain over several kilometres, which would provide unambiguous support for the collaborative human endeavour implied by the major woodland clearances within some catchments shown by the pollen evidence. By later prehistory there were certainly large blocks of cleared land, not least those attested by the establishment of massive Roman temporary camps to hold campaigning armies. Such short-lived sites must have been set on pasture, moor and/or arable, and imply a safe outward view over presumably unforested neighbouring terrain. The largest known Roman camp, at Logie Durno in Aberdeenshire, occupies nearly 150ha. But the plans and characteristics of the field systems that we presently know from most areas of Scotland generally seem to indicate that these are agglomerative and accreted over time, rather than being planned in a single event.

The needs for labour for woodland management, for the extraction of wood for building and heating, the possibilities of feeding pigs on acorns, of ingathering leaf fodder for winter feed for stock as well as woodland's advantages in providing winter shelter, all suggest that well-established later prehistoric farms would have had room for stock, for crops and for woodland in their vicinities. In many parts of the country, charcoal from domestic sites suggests that wood was often used for heating and cooking. Whilst traces of early cultivation systems are found at surprisingly high altitudes in the Borders, it is likely that elsewhere the upland margins of woodland were favoured for transhumant summer grazing of cattle, sheep and goats. Livestock pressure here would have hindered prospects for regeneration at the margins of surviving tracts of upland forest.

WOOD FOR BUILDING

There is an extraordinarily large number of known settlement sites for Iron Age Scotland, and all used wood in their construction. They came in a wide range of structural forms, from individual roundhouses to large enclosed hillfort settlements. Roundhouses had been common in many parts of Scotland since the middle Bronze Age, but it was during the first millennium BC that some of the largest and most elaborate examples were constructed. In broad terms these can be divided into a timber-building tradition predominantly associated with the south and east of the country, and a drystone-building one (which still entailed the use of substantial structural timbers) in the north and west. The latter included the many variants of the Atlantic roundhouse tradition, including spectacular broch towers such as Mousa in Shetland (see Plate 2.2).

In the south and east, Iron Age timber roundhouses are frequent – occasionally isolated and sometimes as components of larger, enclosed settlements. Variant forms have been identified from the nature of the archaeological 'signature' left by their foundations. These were all substantial timber structures, with great conical thatched roofs, and from the outside the visual impact of the different varieties would have been more or less the same.

The basic timber requirements of these buildings can be deduced from excavation evidence, supported by the results of experimental reconstructions, most famously those at Butser in Hampshire. The post-holes dug to support earthfast load-bearing posts inform us of the number and approximate diameter of the original timbers. The remains of burnt posts indicate the species being used, and the re-cutting of features can hint at longevity and refitting of houses. The depths of post-holes can also provide hints of the original heights of the posts. Given that the basic geometry of the roundhouse is fairly simple (essentially a cylinder for the walls topped by a cone for the roof), much of the basic timber framework can be reconstructed from the surviving ground plan. In the reconstruction of an excavated roundhouse from Pimperne in Dorset, it was established that oak trees at least forty years old had been used for the inner ring of posts, with sixty-year-old oaks employed for the more substantial porch posts. Stakes for the outer wall had apparently been obtained from thinnings of

Plate 2.2 *The broch tower of Mousa, Shetland: although most dramatic today for their dry-stone masonry, such structures would have been equally dependent on timber for their internal structure and fittings. (Courtesy of Historic Scotland: Crown copyright reserved.)*

managed woodland some ten to twenty years old. It has been argued that the timber requirements of this structure would have required the careful planting and management of oak woodland over a period significantly longer than a single generation, indeed up to forty or even sixty years before some trees were ready to harvest. Of course, it would have been possible to obtain suitable timbers for a roundhouse in natural woodland, but the numbers and relative standardisation of buildings strongly suggest that at least some of the woodland was actively managed. The resulting roundhouses were in any case exceptionally strong, stable structures, which may have been occupied over several generations if properly maintained. Scotland has its own distinct traditions of timber roundhouse building, but the broad principles, and the levels of demand on local woodland, would have been comparable. Implicit in such architecture, too, is the preparation of suitable timber; building with newly-cut wood might have been avoided.

Houses, of course, are but one component of the timber-building requirements of Iron Age societies. Barns and byres, fencing and enclosing works, would all have made demands on woodland resources. The hundreds of crannogs, artificial timber-built islets found along loch margins, would have provided yet more pressure. Waterlogged timbers from them provide detail of the different kinds of wood used in their

construction: unsurprisingly oak seems to have been important, but alder was also
employed, notably for the piles that supported them. Other timber included hazel;
pieces, still with bark adhering, and with mortice-holes for the wall-frames, were
recovered from Oakbank crannog on Loch Tay. Given the demonstrable density of
settlement in the most favoured Lowland areas of Scotland, local timber supplies
must have been highly prized and carefully overseen: at Oakbank, none of the oaks
seems to have been over a hundred years old, and the alders – probably from the
nearby loch shore – seem standardly to be about forty years old.

In the north and west, where forest cover had been much reduced, the use of
timber resources was even more striking. While the broch towers and other Atlantic
styles of roundhouse are famed for their masterful drystone construction, they also
had considerable implications for timber use in what were by now largely treeless
regions. It is clear that broch towers were roofed buildings, the most elaborate and
monumental representatives of the Atlantic roundhouse tradition which began some
centuries before with simpler drystone-walled structures. The larger, more ambitious
versions would have required ever longer timbers to cover internal floor-spaces some
12m in internal diameter, as well to support upper floors. Some of the timber needed

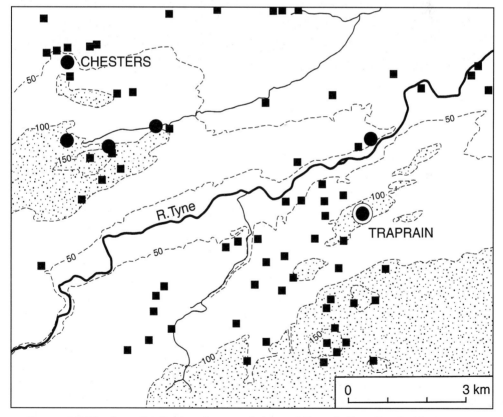

FIGURE 2.2 *The dense distribution of cropmarks around Traprain Law, East Lothian, suggests
high levels of Iron Age population during the pre-Roman Iron Age. (Courtesy of Historic
Scotland: Crown copyright.)*

may have been driftwood (which would have been more plentiful before the much more recent clearance of the great American east-coast forests), and it seems likely that so many large timbers must have necessitated a secure supply, whether accessed locally or through mainland trading partners. Certainly timber in these regions was too scarce to use as the primary fuel source: peat and turf provided heating, cooking and lighting. Indeed, even in the south and east of the country there is evidence of the burning of surface-gathered coal as a domestic fuel, perhaps in response to the local scarcity of timber, on settlements on the East Lothian coastal plain in the immediately pre-Roman period. Nonetheless, wood charcoal is known from numbers of contexts, ranging in size from the substantial central hearths within some buildings in the roundhouse tradition to the microscopic fragments found alongside pollen in peat and lake deposits. This suggests that wood still fuelled much of later prehistoric life, from the preparation of meals to the accompaniment of night-time story-telling.

Broch towers are often considered to have been deliberately flamboyant constructions, statements of power in which the conspicuous consumption of scarce and prized resources such as timber was part of the overall message. Indeed, it may have been the pressures on timber supplies that hastened the demise of these great buildings at the beginning of the first millennium AD; certainly, the structures which succeeded them, such as wheelhouses and cellular buildings, had much more modest timber requirements. Later styles of Iron Age buildings were generally divided into smaller components, or cells, which could be roofed independently by stone corbelling or by using short lengths of timber.

The shift away from the ostentatious use of timber in Atlantic Scotland around the turn of the first millennium seems broadly to coincide with an architectural change identified in the south and east from predominantly timber-built roundhouses to examples with outer walls of stone, although these remained otherwise largely timber constructions. The transition has been identified, for example, at the East Lothian hillfort of Broxmouth, where this new building style may coincide with a weakening of the woodland resource base, or perhaps the imposition of restrictions on access to timber. Population numbers may also have been rising. Control of timber supplies might have been the source of considerable power and wealth in Iron Age Scotland, in societies where so much of one's prestige and status was apparently displayed through domestic architecture, itself dependent on access to plentiful, good-quality timber.

Roman constructions, too, made demands on locally-grown timber, and perhaps also on timber transported from further south. This latter idea, with its implications of yards of stockpiled seasoning wood, has rather fallen from favour, and the likelihood is that most timber for Roman building was locally obtained. Indeed on Trajan's Column, the felling of trees is depicted intimately linked to images of fort-building. In the case of the first-century legionary fortress at Inchtuthil in Perthshire, for example, it has been estimated that the entire fortress required some 30 linear km of timber-framed wood for the walling of its buildings. Both post-hole sizes and nail dimensions – nearly one million nails, many of sizes suitable to assemble major pieces of carpentry – betoken the use of squared timbers of roughly standard sections and

PLATE 2.3 *The experimental burning for a television programme of a timber-laced wall of the type used to enclose hillforts – a process which sometimes led to their vitrification – gives some impression of the requirements for wood both to build and to destroy such fortifications. (Courtesy of James Livingston.)*

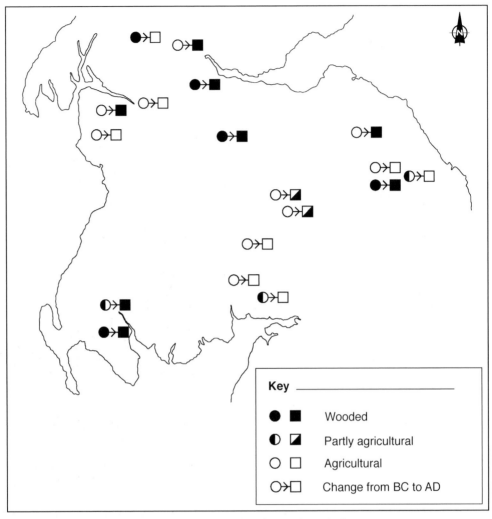

FIGURE 2.3 *Increasing inroads into the woodland cover from the late pre-Roman Iron Age to the Roman period, as shown by pollen analysis. Note that in several cases land cleared before the Romans later reverted to wood. (Redrawn by G. Mudie from data compiled by R. Tipping (1997).)*

sizes. The fortress, along with temporary camps and other related sites on the plateau at Inchtuthil, suggests a cleared area there well in excess of 100ha; other timber may have been brought from the vicinity. Whilst statistics of this kind indicate dramatic localised impacts on woodland, the simple act of felling broadleaved forest will only usually result in long-term clearance if grazing animals stop regrowth occurring. Otherwise, without interference from livestock or burning, woodlands generally recover as they do after natural catastrophes caused by hurricanes.

By the immediate post-Roman period there is less evidence of imposing individual houses across the landscapes of Scotland, although substantial timber halls – such as at Doon Hill, Dunbar – were in all likelihood being constructed at latest by the sixth century AD. During this period, the profligate use of timber is also associated with power in high-status fortifications like those at Burghead in Moray. Timber-laced defensive walls can be traced back well into the first millennium BC at sites like the hillfort at Abernethy in Perthshire. In many cases the only surface evidence for the former use of timbers is the vitrification of the stone components of their enclosing walls, caused by the intense heat generated when the timbers were ignited. The glassy texture of the once-molten stone that characterises such sites testifies to the extensive use of timber within the fabric of Scottish hillforts over much of the country, and not uniquely in the south and east. Imposing when built, the deliberate firing of these major structures must also have been a spectacular intimation of power (see Plate 2.3).

OTHER USES

The use of timber in the Iron Age was not restricted to building. A whole range of day-to-day activities would have demanded a modest yet steady supply of wood. By definition, this is a period in which metallurgy played a part, although iron-working may have had a relatively limited impact prior to the Roman period. Wood and charcoal would have been required for smelting and smithing, even if direct archaeo-logical evidence for these is often elusive.

Timber would have been essential, too, for vehicles used in transport, both by land and water. The recent excavation of a cart (or chariot?) burial at Newbridge, close to Edinburgh, provides a rare glimpse of the sort of two-wheeled vehicle which would have been available at the upper end of the social scale. Such vehicles are rarely deposited in archaeologically recognisable contexts, where the fine wood-working traditions required can begin to be appreciated. Otherwise, instances of preservation are rare. A solid wooden wheel from Blair Drummond Moss, for instance, preserved in the peat, and dated to around 1000 BC, is probably more representative of the simple but functional vehicles that would have characterised later prehistoric Scotland, than the fine-spoked wheels recovered at Newbridge.

As for water transport, dug-out canoes, which may have been used for coastal as well as riverine journeys, form the bulk of the evidence. Iron Age examples include that from Loch Arthur in the Stewartry of Kircudbright. Over 13m long and some 1.5m across at the stern, it bears witness to the great size of trees from which some examples were made. Not all were so grand. Archaeology is otherwise silent on Iron Age sea transport, and it is to the representations on Pictish symbol stones that we must look for indications of the wooden craft that must have a lengthy pedigree in Scottish prehistory.

Archaeological evidence for wooden furniture and internal domestic fittings of Iron Age date is almost wholly lacking. Yet it seems improbable that such great feats of carpentry as the timber roundhouses of southern Scotland or the broch towers of the north and west were not accompanied by appropriate internal fitments. As in more recent periods, wood must have been used for a whole range of domestic purposes which are seldom seen directly in the archaeological record: serving and storage vessels, pails and barrels, handles for knives, weapons and agricultural tools, to name but a few. It may well be that the highly developed domestic pottery tradition of the Western Isles was mirrored by a similarly developed domestic wood-working technology in less treeless parts of the country. Aside from occasional finds of lathe-turned wooden bowls from crannog sites and other wetland locations, however, the evidence remains elusive.

Indications of the use of wood in Iron Age ritual and religion are, perhaps predictably, even more fugitive. The traditional view of 'Celtic' religious practices suggests that the veneration of wooden figures, and indeed sacred trees, was a central part of pagan belief. Classical writers, such as Lucan, were keen to contrast the gory and chaotic natural groves where the Celts worshipped with the ordered world of Mediterranean religious practice. Recently the concept of a pan-European Celtic identity and belief system has come under attack, and it would be unwise to import suspect tales of Celtic barbarism from pre-Roman Gaul to Iron Age Scotland. Nonetheless, there are tantalising glimpses in the Scottish Iron Age of the role of wooden images in pagan religious life. Most impressive of all is the Ballachulish statue, a near life-sized nude female figure, roughly carved in oak with inset eyes of quartz, found by workmen in 1880 (see Plate 2.4). Dated to around 700–500 BC, this image was apparently associated with some form of wicker hut or platform which was unfortunately not recorded in detail. Recent geophysical prospection at Ballachulish Moss, however, has hinted at further timber structures, which may indicate the remains of an Iron Age religious site.

More prosaically, timber would almost certainly have been used in considerable quantities for the cremation of human bodies throughout later prehistory. The evidence for cremation at this time is largely negative, that is, the near total absence of inhumation graves, but it was probably one of the major means of disposing of the dead.

The Wood-working Toolkit

What tools were available to later prehistoric lumberjacks and wood-workers? Axes appear at least from the Neolithic, first in stone and then in copper and its alloys. In later prehistory, and perhaps especially in the Roman period in Scotland, the range of tools used to cut down trees, and thereafter to modify the timber, increases markedly. Although the evidence is fragmentary, it seems that carpentry became increasingly diversified. For instance, production of the Newbridge vehicle would have required specialist wheel-wrights and equipment such as spoke-shaves, or at least draw-knives. Iron Age spoked wooden wheels, fitted with shrunk-on one-piece iron tyres, are certainly indicative of high-quality craftsmanship, and a technology that was subsequently lost for many centuries.

The available evidence consists of the tools themselves, and the surviving timber-work, most remarkably from crannog sites, where waterlogging means that extraordinary detail – for example in the form of toolmarks – can remain readily appreciable. Carbonised timbers, for instance within burnt fortifications, can also provide evidence for mortice and other joints. In terms of the actual tools, the data is skewed by varying

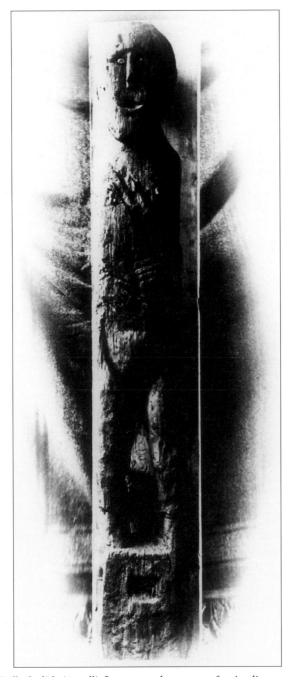

PLATE 2.4 *The Ballachulish (Argyll) figure, seen here soon after its discovery, before shrinkage and decay rendered much of the detail obscure. (Reproduced by permission of the Trustees of the National Museums of Scotland.)*

patterns of discard, the fashion or otherwise for depositing objects in the ground in hoards, and the reworking of scrap metal. Tools are more frequent in late Bronze Age

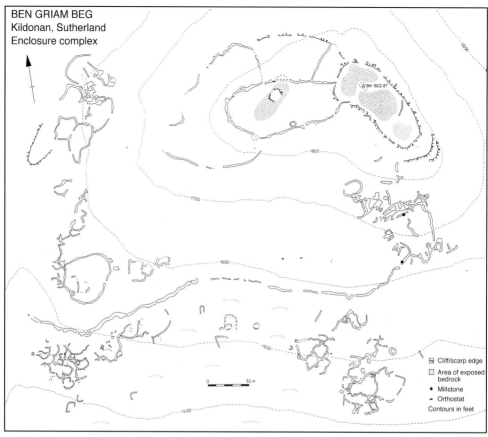

BEN GRIAM BEG
Kildonan, Sutherland
Enclosure complex

Cliff/scarp edge
Area of exposed
bedrock
Millstone
Orthostat
Contours in feet

FIGURE 2.4 *Evidence for later prehistoric field systems provides good evidence in many Scottish landscapes for areas from which woodland had been cleared. The highest and, arguably, most spectacular – albeit undated – system surrounds the summit of Ben Griam Beg, in Sutherland. (Courtesy of R. J. Mercer.)*

and Roman hoards, and on Roman military sites. They are rarer on domestic settlement sites and in Iron Age contexts (see Fig. 2.5).

Axes are numerically the dominant item. Socketed bronze examples of a variety of sizes – some relatively small – were replaced by similar iron ones, and in due course by shaft-hole iron examples. None of the examples used for working wood at the Oakbank crannog, for example, had a blade over 8cm in width. Although stone adzes are recorded in Neolithic times, they seem subsequently to have disappeared from the toolkit until iron ones appear in the Roman Iron Age. In terms of finer working, knives suitable for whittling are known in both copper alloy and in iron. Paring or hollowing out wood was achieved using gouges (with blades curved in cross-section) or chisels; and spoons – and other augers for gouging – and bradawls too were available. Evidence for mortices and half-checks suggests that there were a variety of means of joining pieces of timber. Aside from Roman sites, most famously Inchtuthil, iron nails do not seem to have been in general use for the construction of buildings.

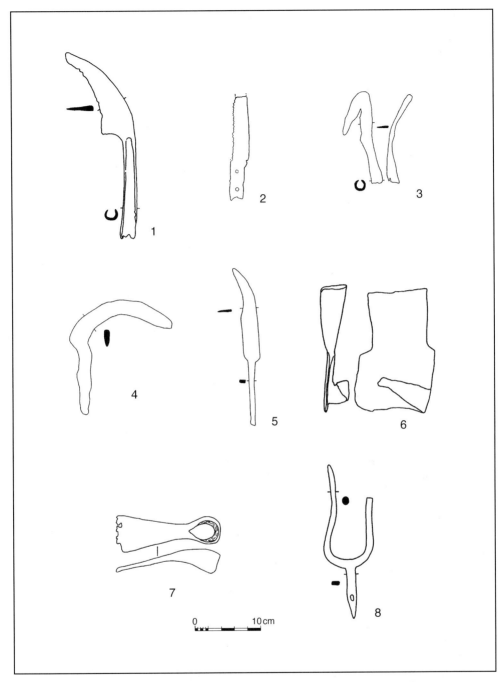

FIGURE 2.5 *During the Iron Age, and more particularly in Roman times, iron tools were increasingly used for tasks related to the working of wood including the pruning of trees, and for the preparation and working of the land from which woodland had been cleared. 1 = bill, Newstead; 2 = saw, Carlingwark Loch; 3 = pruning hook, Newstead; 4 = reaping hook and 5 = knife, both Traprain Law; 6 = peat spade, Blackburn Mill; 7 = hoe, Loudon Hill; 8 = pitchfork, Newstead. (Sources: various, redrawn by George Mudie.)*

The use of small saws, draw-planes and draw-knives (precursors of the later spoke-shave) may all be inferred from the Iron Age, although direct evidence is in short supply. Copper alloy saws are absent in Scotland, and small iron examples may have been used initially for cutting bone and similar commodities rather than wood.

Other items include pruning hooks and saws, probably for fruit trees. Most such items are recorded from Roman contexts much further south in Britain. Both socketed and tanged pruning hooks are known from Roman establishments in Scotland, such as Newstead near Melrose and Camelon, Falkirk, although they are rare. Varieties of bill hooks, useful for splitting branches and withies, making hurdles, and for cutting and laying hedges, are known; socketed examples come from Roman-period contexts at Eckford and Newstead in the Borders. A small pruning saw, of the kind useful for work in an orchard, was also recovered from Newstead.

Throughout our period, too, fire was always in use. In Britain, only coniferous woodland (essentially Scots pine) seems to be susceptible to destruction by natural fire caused, for example, by lightning strikes. The microscopic charcoal found in lake deposits and similar contexts may relate to domestic fires, to industrial practices, to accidental conflagrations on settlement sites, or to the burning of undergrowth and thinnings that must have accompanied woodland clearance. It would be an error to envisage stands of deciduous trees being ignited as a means of clearance, for which, throughout our time-span, the axe must have been the primary instrument. Contrastingly, the burning of heathland, as identified in lowland Fife at Tentsmuir Sands around the middle of the first millennium AD, would not only have encouraged grassland but also inhibited any opportunity for the reappearance of trees. That may have been its most important impact on the wood.

Other Woodland Exploitation

Woodland must have served later prehistoric communities of Scotland in other ways, although clear evidence from archaeological contexts is relatively rare. Surviving forests would have continued to shelter wild fauna and flora attractive to the human population. Among animals, red and roe deer and wild boar probably provided sustenance as well as sport, although it is only after our period – in the Class II Pictish sculpture – that depictions of hunting as an elite occupation appear. On earlier sites, the bones of wild mammals only occur rarely, suggesting that they did not provide a major element of the diet. There are exceptions, such as certain Hebridean sites, although here the deer are unlikely to have been forest animals. Other woodland species would also have been attractive quarry. The greatest of these would have been the bear (and one Caledonian bear certainly crossed Europe to Rome); other fur-bearing mammals would have included beaver, pine marten and polecat. The wolf and the lynx were also present in afforested areas. Woodland birds provided both flesh and feathers.

Acorns and other plant foods would also have provided attractive pannage for domestic pigs. Goats, sheep, cattle and horse are very likely to have contributed to eliminating regeneration both within woodland and especially at its margins. At the early historic fort of Dundurn in Perthshire, in oakwood country, pigs were numerically the second species, but generally cattle and sheep predominate in Scottish faunal assemblages.

Other woodland commodities would have included the range of autumn fruits and berries provided by the understorey plants, as well as hazelnuts. Woodland fungi may have been sought. Resins would have been tapped. Whilst it is likely that most of the diet was sour, the sweetness of honey – by far the most ancient source of sugar – as a foodstuff and as a component of fermented drinks would have been appreciated. Beeswax, too, would have been important, not least in bronze-casting technology. In early times, the discovery of swarms of wild bees in holes in trees must have been especially welcome, and it is likely that bee-keeping, using the honey bee, was already being practised at the start of our period. Especially at the margin of woodland and pasture, with its associated seasonal tapestry of flowers, bees would have been able to exploit both environments. With its cattle herds and flocks, the country may thus have been a land of milk and honey, in which its forest was a significant component.

NOTE

The authors are grateful to Kevin Edwards and Mike Church for assistance.

CHAPTER THREE

Sufficiency to Scarcity:
Medieval Scotland, 500–1600

Anne Crone and Fiona Watson

INTRODUCTION

The Middle Ages is the first period in Scottish history with sufficient evidence to allow collaboration between the archaeologist and the historian. Archaeologists seek to measure past activity scientifically, through the study of excavated materials and evidence such as tree-rings and pollen analysis. Medieval historians, because of the scarcity of the written record in Scotland, deploy their skills in piecing together scattered evidence into plausible hypotheses. The archaeologist therefore provides the bedrock of attested physical evidence while the historian supplies the economic, social and political context.

The period covered by this chapter begins in AD 500, approximately the point at which the Romans, who had had a limited influence on Scotland anyway, left altogether. It ends around 1603, the year in which the Scottish king, James VI, inherited the throne of England and took up permanent residence in the south. By AD 800 a kingdom of Scotland was emerging out of the disparate groups who had arrived over the preceding millennia. In 1603 the political map was redrawn by the new relationship with England. Between those dates, a basic market economy was established, partly replacing a system based on the exchange of goods and services in kind. The habit of writing down important transactions also emerges, allowing us glimpses of landholding structures and the framework governing control over natural resources.

Population fluctuated significantly during the Middle Ages, though we are by no means sure even of approximate round numbers. At the beginning of our period, temperatures had already become milder and, during the eight centuries up to 1300, this allowed numbers to expand, and the cultivation of land that had previously been regarded as too marginal to bother about. Moorland and woodland both retreated, though not to the same extent as in earlier periods of human expansion (see Chapter 2). But timber continued to play a fundamental role in almost every aspect of daily life, most obviously as a building material. Woodlands were also hunting playgrounds for the elite, poaching grounds for others and pasture-land for domestic animals.

THE EARLY MIDDLE AGES:
BRITONS, SCOTS, PICTS, NORSE, ANGLES

The period before 1100 still relies almost entirely on archaeology to show how people utilised wood in their everyday lives. Some of the best-preserved evidence comes from crannogs, artificially-constructed islands found in lochs throughout Scotland. One, at Buiston in Ayrshire, dating from the sixth to seventh century, consisted of a round-house encircled by a timber palisade (see Plate 3.1). The roundhouse was constructed almost entirely of wattle, with double-skinned walls of alder and hazel posts interwoven with hazel withies. There were no internal supports, the walls simply curving inwards to form the roof. The interior of the house was more sophisticated, with a floor of alder planks pegged to oak joists and dressed and decorated panels forming movable partitions. The construction of the palisade is markedly different from the apparently flimsy construction of the house. The earliest was a simple affair of massive squared alder posts but this was replaced by increasingly complex oak structures as the builders tried to provide the crannog with a robust and stable perimeter. The palisades indicate a level of carpentry and access to suitable materials not evident in the house structure. This contrast between monumental external construction, designed to be visually impressive and enduring, and less permanent domestic construction, is seen in many early medieval sites.

By studying the wood remains, it is possible to get an insight into the woods themselves. Alder carr fringed the loch around Buiston, while hazel- and oakwoods were some distance from the settlement. These oakwoods apparently comprised areas of coppice and young maiden trees, growing in clearings created by earlier felling, interspersed with areas of mature standards two or three centuries old. There is no evidence that the woodlands changed much over half a century, suggesting that the crannog-dwellers carefully managed this resource by selective felling throughout a large area of woodland, thus avoiding overexploitation of any one stand.

On Iona, early seventh-century deposits in the enclosure ditch hint at what the famous monastery of St Columba may have looked like. Fragments of decorative panelling and large structural timbers of oak suggest a building of sophisticated design and construction. Adomnan, a seventh-century abbot of Iona and chronicler of Columba's life, describes the transportation of dressed oak and pine to Iona 'for the great house', perhaps a timber-framed building roofed in wood. An illustration of the biblical Temple in the *Book of Kells*, probably written on Iona, shows a building with shingles (wooden roof tiles): fragments of split oak resembling shingles were found in the enclosure ditch, along with large amounts of small roundwood. Adomnan also writes of 'bundles of wattles' being brought by ship to build the guest-house. The construction of substantial timber chapels contrasting with domestic structures built of wattle screens suggests that the former were viewed as monumental buildings, built for long-term survival, while the domestic ranges would be frequently replaced just as at Buiston.

Oak and ash grew on Iona when Columba arrived, according to pollen evidence, but these species were felled and replaced by quick-colonising species like birch and

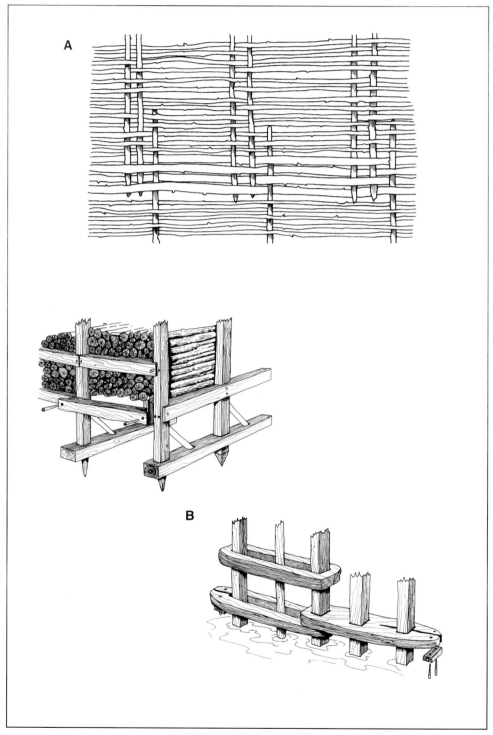

PLATE 3.1 *Reconstruction of (a) the wattle walling and (b) two of the palisades of Buiston crannog, Ayrshire. (Courtesy of Historic Scotland: Crown copyright.)*

willow. The monks subsequently had to get wood for building from the mainland and from Mull. Adomnan records that dressed timbers of pine and oak were 'drawn over land' to Iona and on another occasion that oak timbers were towed by currach from the mouth of the River Shiel to Iona, a distance of some forty miles. He also implies that the monastery had the right to a shipload of wattles growing in the field of a lay-man on either Mull or the mainland.

In the Northern Isles timber supply was also a problem, but once they came under Norse occupation building styles changed, reflecting access to the Scandinavian forests. At Bucquoy, in Orkney, Pictish houses of seventh- to eighth-century date were of 'cellular' design, common on the islands, with thick walls of turf providing support for the roof timbers and removing the need for timber uprights. The short lengths of timber needed (up to 4m) were probably found as driftwood. During the ninth century these buildings were replaced by the characteristic Norse longhouse which required large quantities of timber and upright supports. That the newcomers had access to more plentiful supplies of timber than their predecessors can clearly be seen in a thirteenth-century building at Papa Stour, Shetland, which had a floor of Scots pine planks and traces of benches of oak planking along the walls (see Fig. 3.1). The timber had been debarked and dressed, probably at source in Norway, which appears to have been common practice. Even the birch bark that was found in quantity would have been imported: it was used as roofing insulation under turf, as tapers for lighting, and for making containers. The only species that was locally available throughout the Northern Isles was willow, which the Norse exploited as fuel and for the manufacture of small domestic articles.

The Norse were, of course, renowned for their seamanship. In a small, clinker-built boat found on Sanday in the Orkney islands, the strakes and keel were of oak, with ribs and a washrail of pine. The timber could have come from either Scandinavia or the Scottish mainland, although analysis of sand grains caught in the caulking favours the former. However, Laxdaela's Saga records that, while in Caithness, Aud the Deep-minded 'had a ship built secretly in a forest'. We need not take this too literally, but it indicates how important the timber resources of the north Scottish mainland were to the Norse raiders and settlers. It has been suggested that the Earls of Orkney expand-ed their territorial control as far south as the straths and dales of Easter Ross in a deliberate attempt to access the rich timber resources of this region for the upkeep and maintenance of their naval fleet.

Building in wood rather than stone was a distinctive characteristic of the inhabi-tants of northern Britain, if Bede, the Anglian historian of the eighth century, is to be believed. He states that building in stone was 'a method unusual among the Britons' and records that Nechtan, the eighth-century Pictish king who converted to Christianity, had to send for help from the English to build a stone church. Furthermore, he describes a church on Lindisfarne as built 'after the Irish method [*modus scottorum*], not of stone but of hewn oak'.

Nevertheless, Bede's own race left numerous settlements in southern Scotland built almost entirely of wood. At the Anglian monastery at Hoddom, near Dumfries, a mix-ture of post-and-wattle and post-and-panel construction was employed, the former

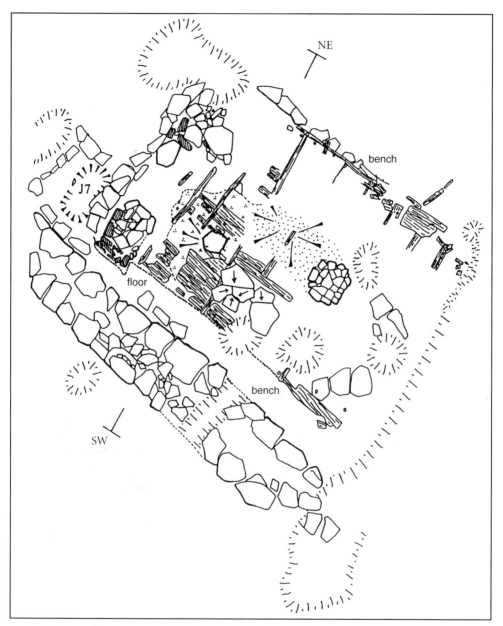

FIGURE 3.1 *The Biggins at Papa Stour, a Norse building showing the wooden floor of pine flanked by benches of oak. (Courtesy of Barbara Crawford: from Society of Antiquaries of Scotland Monograph 15 (Edinburgh, 1999).)*

using a mixture of species while the latter was invariably built of oak. Woodland composition nearby also appears to have remained relatively stable over a period of many centuries, perhaps again implying some sort of management. Large standard oaks continued to be plentiful into the medieval period and young trees or coppice were available which were cropped frequently but perhaps not systematically.

The great wooden hall of the chieftain features prominently in the Anglo-Saxon epic, *Beowulf*: 'a great mead-hall, meant to be a wonder of the world for ever . . . its gables wide and high'. Buildings of this type have been found in southern Scotland and also throughout the Pictish heartland between the Firths of Forth and Moray. One hall on Doon Hill, East Lothian, is probably a result of Anglian expansion in the early seventh century, but it replaced an earlier one, suggesting long, rectangular buildings with porches at each end were also a native type. In this case the building was 23 m long, with walls of post-and-panel construction set into a trench. These were prestigious, monumental structures, using oak.

Large oak timbers are found again in many Pictish fortifications, which often used vast amounts of timber. At Burghead, a promontory fort of seventh- to eighth-century date on the Moray coast, transverse timbers at least 3.6 m long were laid in close-set rows, fixed to longitudinally-set planks using large iron nails. Joinery such as that on the palisades of the crannogs relied on wooden trenails, so the quantities of iron nails found at Burghead, and also at the Pictish citadel at Dundurn in Perthshire, suggest high status.

We have little evidence for the houses in which the peasantry lived, presumably because their flimsy buildings have left scant trace. At Easter Kinnear in Fife, there is one such building, a scooped semi-subterranean dwelling with wattle and daub walls of hazel and oak with a little alder, birch and willow.

For the ninth to eleventh centuries, evidence is sparse, but hints as to what some contemporary buildings may have looked like come from an unusual source. A distinctive group of tombstones known as hogbacks, found scattered along the east coast from the Tweed as far north as Brechin, depict buildings with carved finials, wattle panels for walling, and wooden roof shingles (see Plate 3.2). These tombstones have

PLATE 3.2 *Hogback stone (no. 5) from Govan parish kirk: such tombs appear to illustrate long-houses with shingle roofs, of a Scandinavian type.* (© *The Trustees of the National Museums of Scotland.*)

a Scandinavian origin and may nostalgically depict buildings from the homeland, but examples of longhouses with the distinctive bowed sides have been excavated in Scotland. The earliest structure recorded in Inverness is a plank-built wall dated to the late twelfth century, of a style harking back to earlier Norse traditions.

USING AND MANAGING THE RESOURCE, 1100–1400

In the twelfth century, many nascent towns were given burgh status, developing rapidly into the administrative, industrial and trading centres that still exist today. There is little physical evidence for their earliest development, possibly because the buildings within the towns were so flimsy. Dendrochronological analysis of structural timbers from excavations in Perth, Aberdeen, Inverness and Glasgow has identified a major phase of building activity in the late twelfth or early thirteenth centuries (see Fig. 3.2).

Between the twelfth and fourteenth centuries most urban buildings for which there is archaeological evidence were single-storey buildings of post-and-wattle, built in the backlands of burgage plots which extended away from the frontages lining the market place (see Plate 3.3). These are clearly the kind of simple structures that the people of Lanark were describing when, after a fire had burned down the town in 1244, they asserted 'that with six or eight stakes they would soon have new houses'.

Such structures are characterised by the absence of elaborate carpentry and the abundant use of relatively small undressed roundwood. The withies used to build the wattle walls of a group of buildings excavated at Kirk Close, in Perth, were half hazel

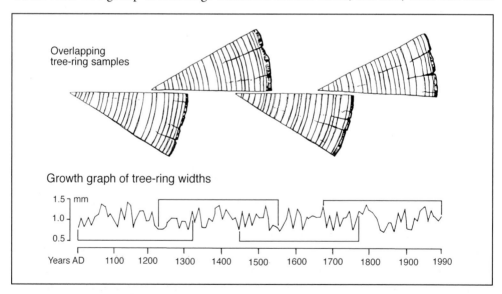

FIGURE 3.2 *How dendrochronology works. Every year a tree lays down a certain amount of wood during the growing season, and each year has a unique pattern. By gathering a series of overlapping tree-ring samples, a chronology can be built up – in this case for oak – of patterns that go back for hundreds, even thousands of years (After Ian Tyers, 'Tree-ring dating', Medieval World, 2 (1991), pp. 12–18.)*

PLATE 3.3 *Building a house in medieval Aberdeen: basically wattle and poles. From an archaeological reconstruction. (Aberdeen Art Gallery and Museume.)*

and 17 per cent alder, together with small amounts of birch, willow, elm and fruit trees. This composition does not necessarily reflect the use of managed coppice; the hazel and alder may have come from such a source but the remaining species suggests the gleaning of hedgerows and woodlands around the town. Royal grants often gave the burgesses rights to gather timber and fuel in the king's forests. However, not everyone who lived in a town automatically became a burgess (there was a property qualification), which begs the question as to what everybody else did for supplies.

Small amounts of dressed timber – primarily oak – were used in these buildings, all of which displayed evidence of re-use. Grander, better-built buildings probably lay along the frontages of the plots and the re-used oak timbers may have been scavenged from the demolition of these more prestigious structures. Unfortunately, we have little archaeological evidence for the type of buildings that lined the frontages, mainly because these were the areas where redevelopment was most intensive and earlier structures have consequently not survived.

The evidence for rural settlement throughout this period is equally scant, presumably because the bulk of the population continued to live in structures that left little imprint on the ground. The late thirteenth- or early fourteenth-century byre-dwellings excavated at Springwood Park, Kelso, were of cruck-frame construction, where pairs of timbers rest on the ground or on a stone foundation and meet at the top in the shape of a wishbone (see Plate 3.4). Crucks probably formed the framework of most rural buildings throughout the medieval period. At Springwood Park, they may have been elm and ash, the major components of the charcoal from the site. There was

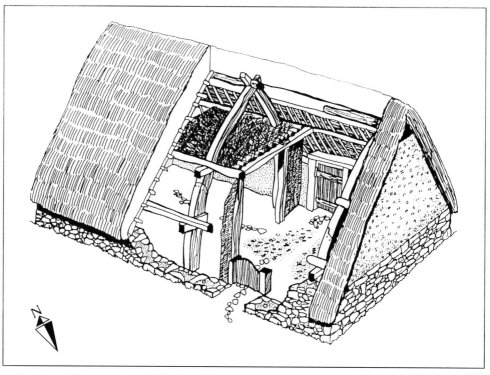

PLATE 3.4 *Reconstruction of the late thirteenth- or early fourteenth-century byre-dwelling at Springwood Park, Kelso. The three pairs of wooden crucks rested on padstone foundations. (Courtesy of Historic Scotland: Crown copyright.)*

noticeably little oak in the charcoal. As in the towns, oak was probably not readily available to the peasantry.

The extension of Norman influence into Scotland in the twelfth century was accompanied by the construction of impressive wooden motte-and-bailey castles by both crown and nobility, particularly in north-east and south-west Scotland. At Castlehill of Strachan, Aberdeenshire, built c. 1250, the type of wood used in the construction suggests that oak was not always available even to the rich and power-ful. Here, on top of the high earthen motte, a circular hall with walls of alder wattle stood within a timber palisade mainly of alder and hawthorn, the spininess of the latter making it particularly suitable for use in the defensive perimeter.

Timber castles came to be replaced by stone ones, but the most impressive medieval buildings were not built by nobles or even kings but by the Church. Stone, of course, was also the basic material of their walls and towering arches, but the spans of the great roofs imply the use and availability of large timbers, and throughout the thirteenth and fourteenth centuries there are numerous records of grants of oaks to religious bodies. Thanks to the English raids and to the Reformation, original timber structures rarely survive in the Scottish abbeys but at Jedburgh one corner of the cloister was built over a raft of huge oak beams, one 9.5m long and 0.70m wide, which had been felled sometime between 1235 and 1271.

Scotland's woodlands during the Middle Ages were subject to a complex system of control. That control depended most particularly on who owned them and, therefore, the legal framework pertaining to that ownership. The Latin term in *libera foresta* (in free forest) is the most fundamental legal definition from this period and encompasses a very specific set of regulations.

If forest status was granted, this gave absolute control, theoretically speaking, to the owner of everything in, on or above the ground under the terms 'vert and venison'. The *vert* (literally, green) referred to the living plant life present within a forest, including the timber (both greenwood and dead wood) and also other materials, such as broom, which could be used as a roofing material, especially by the peasantry. Venison, of course, referred to the all-important primary resource of the forest, the deer, though other creatures, particularly birds, were also hunted.

The main role of the forest, therefore, was to act as a place for hunting. It is only in the more recent past that the word has come to mean the tracts of trees which, we presume, were usually found in the vicinity to some extent. However, Scotland contains a number of forests which, so far as we know at present, have not seen trees since before the historic period.

Hunting was a vital element of life if you were a member of the upper echelons of society. Access to hunting was one of the most obvious ways in which early society expressed its increasing stratification. The forest laws brought to Britain by the Normans are the first codification of rules governing access to the resources of the area designated as forest. However, these laws almost certainly reflected practices and regulations that had already existed for many centuries.

A number of officers managed the system. The official forester was usually a major landowner who was given the job as a mark of royal favour. Under him was the actual forester on the ground and a series of under-foresters, who were allowed remission of rent or free access to timber as an incentive to perform their job. Their evidence of contraventions of the rules would have been crucial to the running of the forest courts, held with varying degrees of regularity throughout the land.

The usual perception of forests, created in part by the story of Robin Hood, is of large tracts of royal land out-of-bounds to the majority of the population, which nevertheless yearned for access to them and their resources. That image may or may not be true for England, where central authority was one of the strongest in Europe. In Scotland, exclusion from the forests seems to have been less stringently applied (certainly owners allowed use, including grazing, for a fee). However, this should not lead us to conclude that legal ownership was in any sense less well established in law in Scotland; it merely means that the exercise of ownership might be different.

Essentially, Scottish forest regulations were about managing competing priorities among different levels of society – the grazing of animals; the use of timber and other forest resources; and the need to maintain the deer and bird populations, as well as the environment in which they lived, for hunting. Owners could also be users in other people's forests, so they understood perfectly well the need for give and take. This was particularly true for the inhabitants of the various abbeys and monasteries who were granted forests for their own use, but often had the right to graze their animals or extract timber (especially for building purposes) from forests and woods all over the country.

The whole issue of rights, legal or traditional, to various aspects of forests, woods and trees, was a minefield of contradiction, not least between theory and practice. The complexity of the situation is illustrated by a court case of 1256. The monks of Lindores were forced to take one of the Earl of Strathearn's estate officers to court to gain access to their legitimate supply of timber from the wood of Glenlichorn in that earldom. The monks were fortunate: their rights to take timber were written down, as is usually the case for such an ecclesiastical organisation, but the earl's officer remained unhappy about this access to the timber, presumably because he believed that it belonged categorically to his master. He only grudgingly admitted the monks' right and insisted that he or his officer should be informed when any wood was cut in future. Such an attitude seems to have more in common with the eighteenth century than the thirteenth and should remind us not to be too hasty in drawing conclusions just because the evidence is thin on the ground. There was plenty of control and management in the Middle Ages – it is just that we do not always have the evidence to say very much about it.

So far as we can tell, there was no uniform legal right of access to timber or the woods as a whole. However, in reality tenants must have had some sort of agreement with the owner, covert or overt, whereby they could get what they needed. They might have had to pay for it or they might have been given an annual amount, to be cut within sight of the owner's officers.

Grazing rights were another tricky issue which fluctuated over time, depending on economic circumstances. Given the fact that Scotland's main exports consisted of animal products such as wool and hides, the use of woods for shelter and grazing was a fundamental part of the agricultural routine. In the Highlands many such rights were traditional and, in this predominantly oral culture, usually not written down. We only find out much later, when circumstances dictated that the landowners wanted rid of them.

SIGNS OF SCARCITY, 1400–1603

Parliament does not seem to have particularly concerned itself with trees until quite late on (in comparison with salmon, for example). The main reason for the sudden interest may be a perception that, by the fifteenth century, large timber in the Lowlands was in short supply and something needed to be done to reverse that situation. The first known laws on timber management came in 1424, when James I returned home from imprisonment in England, prompting a flurry of legislative activity.

Legislation encompassed acts both banning the general use of wood without permission, while actively encouraging planting. As ever, the constant reiteration of legislation makes it clear that, like so many government initiatives, it did not necessarily meet with the approval or the all-important co-operation of the wider public. The penalty for contravention of these regulations was sometimes raised even to the death penalty for a third offence, but over the period as a whole, such draconian measures had little effect. The key to the effectiveness in the regulations lay with the usual kinds of economic and social forces at work on the population as a whole.

The surviving documentation also indicates that distinctions were already made between enclosed and unenclosed woods, although the physical evidence for enclosure is scarce. Sophisticated systems of management of the timber itself, including division into haggs (felling coups), were used by some religious establishments. The extensive system of woodbanks that criss-cross Bowden Moor in the Borders is thought to have been constructed c. 1500, probably by the monks of nearby Melrose. Unfortunately, it is not known whether woodbanks elsewhere, such as those at Garscadden wood north of Glasgow, belong to this period or to the later period of intensive coppice management in the eighteenth century. The quantities of withies needed to build even the simplest of medieval buildings imply that coppiced woodland must have existed, but what scant evidence there is suggests that cropping may have been carried out on a casual, opportunistic basis rather than the more formal organisation that emerged in England. Yet the Church in particular (but also the higher echelons of secular society) operated across international boundaries, and knowledge was easily transferred across Europe. So it is likely that the manager of a monastic estate in Scotland would have had the opportunity to learn (directly or indirectly) of advanced wood management from England or France and then be able to pass knowledge on to others.

Even within the monastic domain, however, good woodland management might give way under other pressures on land use. The monastic house of Coupar Angus in the fifteenth and sixteenth centuries managed woodland at Campsie by dividing it into quarters, maintaining enclosure and appointing local tenants to have the responsibility of being foresters. However, the demand for grazing land was such that animals were allowed into the enclosures, and part of the wood was eventually reported as 'decayed'. Two centuries later, there was no trace of it.

The bulk of surviving buildings after 1400 are stone-built structures, prestigious castles and churches in which timber would have provided the roof, the flooring and the internal fittings. Of the original timber components few retain much more than the roof structure and what timber survives is invariably oak. Throughout our period this was the species most favoured in high-status buildings because of its strength and durability.

The most spectacular surviving timber roof is the hammerbeam roof at Darnaway Castle in Moray (see Plates 3.5 and 3.6). Originally thought to be late fifteenth century, dendrochronological analysis demonstrated that the oak trees used to build the roof were felled in 1387, making it the earliest surviving example of this type of roof structure in the British Isles. Original ceiling structures also survive in many of the royal residences (Holyrood, Falkland, Stirling and Edinburgh) and in some of the more important ecclesiastical buildings, such as King's College Chapel, Aberdeen. However, we also should not forget the importance of other forms of interior decoration and the extent to which working in wood was very much an art form. King's College Chapel, for example, retains its original wooden furnishings, including the canopied

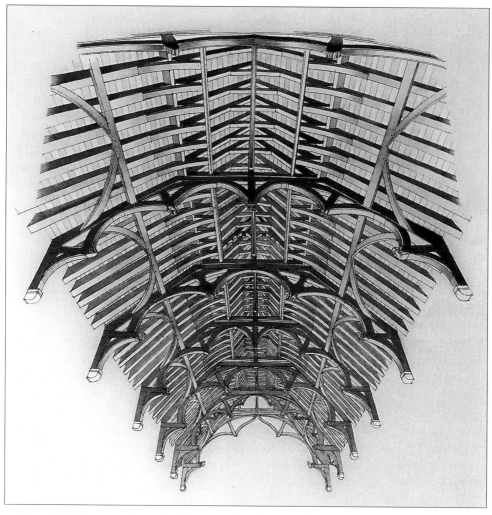

PLATE 3.5 *The great oaken roof at Darnaway Castle, Moray, dated to 1387 by dendrochronology, the earliest hammerbeam construction in the British Isles. (Royal Commission on the Ancient and Historical Monuments of Scotland: Crown copyright.)*

choir stalls and screen lavishly carved with openwork tracery designs, though these, Flemish in design and concept, may have been imported already worked.

The stone skeletons of that most ubiquitous of building types in Scotland during the later Middle Ages, the tower-house, litter the Scottish countryside. Very few retain much of their original woodwork, though occasionally floor joists survive; only two, Alloa Tower and Bardowie Castle in Stirlingshire, retain an intact roof structure, again of oak. A very unusual *in situ* survival is the seat over the garderobe shute (toilet) at Stoneypath Castle, East Lothian, where a thick chord trimmed off an oak log has been used.

One apparent gap in architectural development in Scotland, in contrast to England, France and Germany, is the relative lack of a widespread or well-developed timber

PLATE 3.6 *Carvings on the Darnaway roof: (opposite) a king and (below) a wolf-slayer. (Royal Commission on the Ancient and Historical Monuments of Scotland: Crown copyright.)*

box-frame building tradition. In England completely timber-framed buildings were being constructed up until the eighteenth century and are a familiar feature of both rural and urban landscapes. There are no known timber-framed buildings of this type in the Scottish countryside, and within the towns the only known examples are late developments and only partially timber-framed.

PLATE 3.7 *Oaks in the medieval hunting park at Cadzow, Hamilton, Lanarkshire. These trees have been dated at least to the fifteenth century, and grow on the ridge-and-furrow of still earlier fields emparked to provide a lord's hunting. This tree was photographed in the 1870s, but is still there. (Courtesy of the Royal Scottish Forestry Society.)*

None of the excavated evidence for buildings in the early burghs reveals any antecedents for the development of timber-framing, although the post-and-sillbeam construction seen on the High Street, Perth, could perhaps be seen as such. However, building in timber is considered to have been the norm in the towns until the second half of the sixteenth century and urban stone buildings appear to have been unusual enough to draw comment. This is surprising given the apparent difficulty of obtaining local timber of good quality and the increasing reliance on foreign imports that becomes apparent towards the end of the fifteenth century.

Unfortunately, the evidence needed to track the development from timber to stone building in towns is not presently available, although it may lie hidden within buildings of later periods. The wooden roof of a seventeenth-century house in the High Street, Brechin, revealed remnants of a timber-framed building, dendrodated to 1470, that had been re-used within the later building. The timbers showed redundant mortice- and peg-holes characteristic of a box-framed construction.

The earliest known illustration of Scottish urban housing is the view of Stirling Castle with the burgh below in a fifteenth-century manuscript of an early history of Scotland, the *Scotichronicon*. It depicts jettied buildings (the upper storey projects out over the ground floor), but whether these are entirely timber-built or a mixture of stone and timber is not evident. The only surviving examples of timber-framing date from the late sixteenth century and these are built mainly in stone with timber-clad frontages. John Knox's House, in Edinburgh, c. 1570, is entirely of stone with a jettied frontage of timber. Kinnoul's Lodging, in Perth, dated to c. 1600 and dismantled in the 1960s, had a timber-framed frontage built over a masonry wall at ground level, while the other three load-bearing walls were entirely of stone (see Plates 3.8 and 3.9). This seems to have been a peculiarly Scottish fashion, a visitor to Edinburgh in 1598 commenting on buildings 'faced with wooden galleries, built upon the second storey of the houses'. The only almost wholly timber-framed house recorded to date in Scotland is a house in the Lawnmarket, Edinburgh, which was erected around 1580 and demolished in the late nineteenth century (see Plate 3.10). The profligate use of so much timber in a building of this late date may have been an imitation of English and Continental fashions rather than a continuation in the Scottish building tradition.

Records of fires in many Scottish burghs throughout the seventeenth century and the consequent attempts of burgh councils to enforce building in stone show that timber continued to be a major component in urban buildings. Even roof coverings were sometimes of wood. The roof of the Canongate Tolbooth, Edinburgh, built in 1591, was covered in oak shingles. Documentary references suggest that this was the preferred roof covering for many high-status buildings until slate began to be quarried.

We know that small amounts of timber and wood products were being imported during the fourteenth century. Tree-ring analysis has identified fourteenth-century oak of eastern Baltic origin in Queen Mary's House, St Andrews. Northern Poland has been specifically identified as the source of the oak used in a barrel of late fourteenth-century date found in the Gallowgate, Aberdeen. Whether the latter arrived as a barrel containing some more valuable import such as honey or beer, as prepared staves, or simply as timber is not known.

By the late fifteenth century when records of royal purchases become available in the accounts of the Lord High Treasurer and the Master of Works, it is clear that the eastern Baltic was already a major supplier of one particular type of timber. The accounts invariably refer to Baltic timber as 'Eastland boards', indicating that it was imported ready prepared as cleft planking. As time passed, such imports became routine. We know, for example, that of 122 Scottish boats coming from Königsberg in East Prussia between 1588 and 1602, 80 per cent carried wooden boards, often as a minor part of a mixed cargo. The royal accounts often specify that 'Eastland board' was intended for doors, windows and panelled ceilings – woodwork where relatively thin boards were required. A number of examples have been identified by dendrochronology. For example, the source of the panels of the late fifteenth-century Guthrie Aisle painted ceiling and carved tracery panels from Perth was eastern Europe, possibly the modern states of Ukraine and Belarus (see Plate 3.11). The oak of these panels was fine-grained and slow-grown, qualities which are necessary to produce panelling which does not warp with seasoning and which were probably no longer available in locally-grown woods.

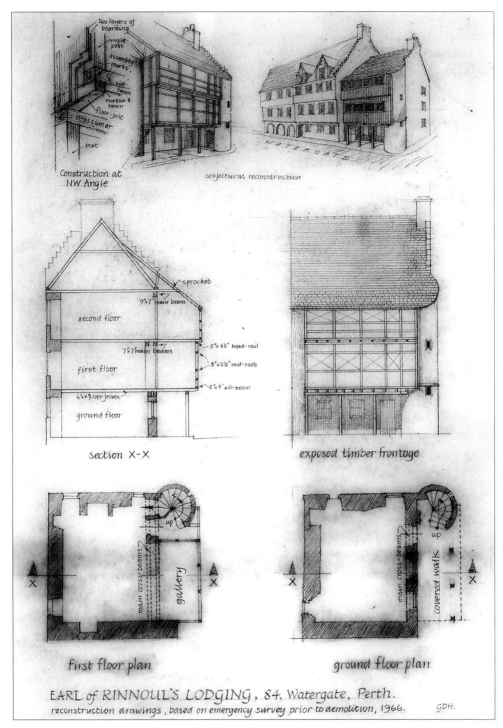

Construction at NW Angle

conjectural reconstruction

WATERGATE

two layers of boarding
angle post
assembly marks
sill-beam
mortise & tenon
floor-joist
Bressumer
post

sprocket
9"x7" main beam
second floor
7"x7" main beams
5"x4½" head-rail
3"x2½" mid-rails
first floor
5"x5" sill-beam
6"x4" floor joists
ground floor

Section X-X

exposed timber frontage

up
main cross-beams
gallery
X X

first floor plan

up
main cross-beams
covered walk
X X

ground floor plan

EARL of KINNOUL'S LODGING, 84, Watergate, Perth.
reconstruction drawings, based on emergency survey prior to demolition, 1966.   GDH.

PLATE 3.8 *Kinnoul's Lodging Perth, c. 1600, as surveyed prior to its destruction in 1966: timber-framed frontage over a masonry wall, characteristic of townhouses of the time. (Royal Commission on the Ancient and Historical Monuments of Scotland: Crown copyright.)*

PLATE 3.9 *Kinnoul's Lodging, Perth, photographed before its destruction. (Royal Commission on the Ancient and Historical Monuments of Scotland: Crown copyright.)*

Other parts of Europe also supplied Scotland with timber, particularly Scandinavia. The royal accounts very occasionally mention Scandinavian timber, and from the mid-sixteenth century both the customs books of the Stavanger region and the Dundee burgh records testify to a considerable volume of timber brought in from Norwegian and Swedish ports. Indeed, the Scottish demand for building oak was so great that for a period the crown of Denmark–Norway prohibited its sale to foreign traders, to conserve its own resources. This timber has now been identified in a number of buildings including the Great Hall at Edinburgh Castle (1511) and the Royal Palace at Stirling Castle (1500 and 1538). Scandinavian timber was also used in 1470 to construct the timber-framed building in Brechin, described above. The timber used in these buildings was oak, but sixteenth-century records indicate that pine, and perhaps spruce, was also imported from Scandinavia in very much greater quantities, although these species have not been identified in any standing buildings

PLATE 3.10 *The only known almost wholly timber-framed house in Scotland, built around 1580 and destroyed late in the nineteenth century: in the Lawnmarket, Edinburgh. (Royal Commission on the Ancient and Historical Monuments of Scotland: Crown copyright.)*

of that date. Scandinavia was clearly the main supplier of standard timber needed for most types of joinery.

Timber was also imported from Ireland to east coast ports such as Ayr and Wigton, but it was very limited in comparison with the Baltic and Scandinavian trade. Southwest Scotland probably had reserves of native oak for longer than other parts of Scotland and so the pressure to import was less. Throughout this period timber from the king's own reserves was still able to supply some of the needs of the royal builders. Oak was brought from Lochaber, from Darnaway in Moray, from the Torwood near Stirling, from Callander, Perth and Clackmannan for use in the palaces at Falkland, Linlithgow and Holyrood, while birch from Callander was used to make scaffolding and trestles.

STATE OF THE WOODLAND RESOURCE IN MEDIEVAL SCOTLAND

Archaeology, palynology and documentary evidence can provide at best 'snapshots' of the state of the woodland resource at particular points in time. Study of the timber itself can yield particular insights, analysis of the tree-ring patterns telling us where it

PLATE 3.11 *Four of the tracery panels found in a house in the Skinnergate, Perth, made of eastern European oak in the early sixteenth century. 'Eastland boards' were commonly imported into Scotland at the time, and the close-grained wood made such fine carving possible. The dimensions of each panel are roughly 1 foot x 1½ feet. (Courtesy of Perth Museum and Art Gallery, Perth and Kinross Council.)*

came from, the conditions in which it was growing and the subsequent quality of the timber. Unfortunately, however, this material evidence is only rarely available. It is as well to remember that woodland history, just like any other, is full of fluctuations, with woodlands re-establishing and spreading as well as being cut down, and many of our 'snapshots' may simply reflect a very local situation rather than a general trend.

At the very beginning of our period the Lowlands were probably still comparatively well wooded, at least in places. Palynological evidence suggests that there had been large-scale land clearance in the preceding Iron Age but that the early historic period saw reduced agricultural activity which allowed scrub woodland to regenerate, though these trends may not have happened everywhere. North of the Forth, the frequency with which woodland terms occur in Pictish placenames hints at a landscape where tracts of woodland were still common. Some archaeological evidence from the early half of the period suggests that local communities, as at Buiston and Hoddom, may have managed their woodland to ensure a continuous supply. Nevertheless, an increasing population would inevitably have contributed to the erosion of woodland cover.

There is some evidence that occasionally this process was reversed. Most of the trees felled for building in the early burghs began growing in the mid-tenth century, suggesting that some woodlands were able to regenerate at about this time. Perhaps some event, such as an epidemic or Viking raids, occurred at around that time, causing human settlement to retract and thus enabling woodland to regenerate. The regeneration of the woodland meant that there were plentiful supplies of local timber available for building within the burghs in the late twelfth and early thirteenth centuries. In some areas, such as Moray, high-quality oak continued to be available throughout the fourteenth century. For instance, timber used in Elgin in 1301 came from slow-grown oaks, some at least 358 years old at the time of felling. This quality of timber was still available to build the hammerbeam roof at nearby Darnaway Castle in 1387 (where one tree was at least 418 years old when felled), and was also used for building at Stirling Castle in the first half of the fifteenth century. It seems that the Royal Forest of Darnaway, the most likely source of the timber in both structures, was able to maintain a stock of exceedingly long-lived trees well after 1400.

Buildings in south-west Scotland also used native oak until relatively late, implying that local supplies were still readily available. By the late fifteenth century builders on the east coast were relying on Baltic oak for panelling, but at the same time painted choir stalls could still be made from boards of native oak, as at Lincluden College, Dumfries. As late as 1590 native oak was being used in the construction of the tower-house at Castle of Park, Kirkcudbright, despite the fact that by this date high-status buildings on the east coast were consistently built of imported oak.

Elsewhere, however, there are indications that the strain on timber resources was beginning to show as early as the fourteenth century. Imported building timber appears in St Andrews, although that could simply be the result of effective marketing by foreign merchants. However, excavations at the deserted medieval burgh of Rattray in Aberdeenshire revealed that the thirteenth-century buildings were predominantly of wood, whereas those built in the fourteenth century employed stone and clay, an indication perhaps of exhausted wood supplies. Certainly an early decline in suitable building timber is often mooted as the reason that a sophisticated timber-framed building tradition never developed in Scotland.

Another important consideration in assessing the likely state of the woods is fluctuations in population. Scotland's population had been growing for some time before the outbreak of the Black Death in 1349, bringing pressure on the remaining woodlands as more land was brought under cultivation to feed the increasing numbers.

The subsequent population reduction should have brought relief to the hard-pressed woodlands in some areas, but, ironically, more complaints of scarcity are heard thereafter than before. Perhaps the decline in cultivation pressure was not mirrored by any decline in grazing pressure.

We must not take too literally the assertion in the Parliament of 1505 that 'the wood of Scotland is utterly destroyed'. In the first place, there is plenty of evidence from the seventeenth and eighteenth centuries that parts of the Highlands were still comparatively well wooded. Second, even the Lowlands had numerous patches of woodland that have often remained under trees to the present day. Nevertheless, it is clear that the Lowlands were suffering from a shortage of substantial timber by the late fifteenth century and in 1504 Parliament ordained that every lord shall plant 'at the leist ane aker of wood' to combat the problem. While the royal accounts occasionally record that 'gret akyn treis' could still be had from the Torwood and Callander woods in 1534–5, more entries record the use of Scottish timber only for laths and sarking, for which only small wood is required. Certainly, the few surviving examples of locally grown oak of this date are young and fast-grown. The native oak used in the roof at Alloa Tower was no more than forty to fifty years old, while timber felled in 1538 and used in the construction of the King's Bedchamber at Stirling Castle was between fifty and eighty years old.

The years from 500 to 1603 (more than a millennium) saw major changes in the nature and extent of Scotland's woodlands. Between 500 and 1350 Scotland had sufficient resources to be able to supply its own requirements for building purposes. Over the next 250 years the Lowlands at least became gradually reliant on imported timber for substantial buildings, and for construction in towns. In the country, local supplies probably still sufficed.

Despite significant changes in the condition and extent of the woodland, some aspects of the relationship between people and woods remained the same. Access to woodland was controlled by the rich and powerful and the produce of the woods was used to reflect their status. Oak, the 'noblest of trees', was very much the preserve of the more wealthy, and was used in the construction and decoration of monumental buildings. Nevertheless, there were many ways in which the peasantry could gain unofficial access to this timber, especially the further away from the source of power (such as the lord's castle) they happened to be. Furthermore, it should not be forgotten that even if the use of timber was restricted (and this was a variable constraint), Scotland's remaining woodlands were extremely valuable to them as a source of shelter for their livestock. Though the people of medieval Scotland had the ability to manage their woodlands successfully, they could not be expected to put the needs of the trees above their own.

CHAPTER FOUR

Using the Woods, 1600–1850
(1) The Community Resource

MAIRI STEWART

INTRODUCTION

As earlier chapters have shown, the great mass of Scotland's woodland cover had already been lost by 1600. It is difficult to assess with any degree of accuracy how much was left, as documentary evidence is scarce, but from the seventeenth century onwards progressively more written records of Scottish life were made and many have survived. These original sources, particularly the records of the great landed Scottish nobles like Buccleuch and Breadalbane, are indispensable to historians. They provide key pieces of the jigsaw, which when matched with old maps, published accounts of travellers, ministers and soldiers, come together to create a progressively more accurate picture of Scotland's woods.

What were the woods really like in 1600? How much was there, where were they, how were they being used by the people and how did this use change them? Sir Walter Scott famously described a Highland landscape as, 'so wondrous wild, the whole might seem the scenery of a fairy dream', but that romantic vision conceals the fact that the Highlands were, as they are today, a working landscape. They change as the economy changes, and their beauty and wildlife are moulded by human action as well as by nature. This has been true of the woods, as of the moors, and of course it has always been true of the rural Lowlands.

It has been said that Scotland's semi-natural woods, today, are not markedly different to what existed 250 years ago, in extent at least. Can we, however, really believe that the semi-natural woods around us, like the great Caledonian pinewoods of Rothiemurchus and Abernethy, or the extensive and uniquely rich Atlantic oakwoods, like those of Taynish or Glen Nant, are really like the woods that our ancestors knew 250 years ago?

Perhaps this is easier to believe when standing among the ancient warrior pines of Loch Arkaig, some of which may already have been well grown in 1746 when the Jacobites were being flushed out from their wooded sanctuaries by Hanoverian fire. As the authors of the pioneering study *The Native Pinewoods of Scotland* said of such pinewood remnants, 'To stand in them is to feel the past.' But then again, it is not difficult, particularly in the Highlands, to find yourself in a landscape, desolate and wild, where a huddle of hardy birch or a solitary rowan cling to the crag or burn side, or a few granny pines are scattered across the hillside, as in Glen Falloch in Perthshire or Glen Derry in the Cairngorms. Did these sorry-looking remnants cling to the seventeenth- or eighteenth-century Scottish landscape just as they do today? Indeed, if we found ourselves standing on the banks of Loch an Eilean in the heart of Rothiemurchus forest 250 years ago, would we feel that time had changed nothing?

In uncovering the story of what befell our native woods after 1600, we must consider wider developments in Scotland's society and economy, for these were times of great change in all aspects of Scottish life: a union of crowns (1603) and of parliaments (1707) and subsequent political upheaval, famines, rebellions, the agricultural revolution and Highland clearances, the transformation of towns, not least Edinburgh and Glasgow, new roads, railways, steamships. All of these, and many more changes besides, had taken place in Scotland by 1850. By then, the influences of the Industrial Revolution and the ascendancy of capitalism were beginning to grip the heart of Scottish life, with far-reaching consequences for both people and woods.

Woodlands in the Early Seventeenth Century

Inevitably, the focus is on Highland woods when describing Scotland's forest history from 1600, because then as today, most remaining 'semi-natural' woods were in the Highlands. The much earlier demise of woodland in the Lowlands has already been charted. But to ignore Lowland woods would leave us with an imperfect picture, if only because management techniques long applied there became more influential in the Highlands during this period. Their very scarcity also meant they would be treated differently.

By 1750, when General Roy was making his wonderful military survey of Scotland and locating the woods for us in the first systematic map of the countryside, we can calculate that less than 2 per cent of the Lowlands supported semi-natural woodland (see Plate 4.2). This distribution was patchy, with concentrations along river valleys, such as the upper Clyde, along Ayrshire rivers and in parts of the Borders, Dumfries and Galloway (e.g. Ettrick and Eskdale). Even in highly fertile areas such as the Lothians, woods remained, like Pressmennan, Roslin, Dalkeith. Indeed all still exist today, as do Mugdock and Garscadden woods, despite their proximity to Glasgow, the latter on the edge of Drumchapel. We know many of these woods existed and were well used in the seventeenth century through written records such as estate papers or through the central Register of Deeds recording wood sale contracts. Around 1750 the overall woodland cover of Scotland was about 4 per cent.

The earliest maps of Scotland originated, however, a century and a half before Roy, and they show a similar pattern, though one with rather more wood. Timothy Pont was an enigmatic and intrepid 'son of the manse', who between around 1585 and 1596, travelled 'a' the airts' of Scotland, sketching and noting down what he found (see Plate 4.1). From his rough maps, a Dutch mapmaker, John Blaeu, published in 1654 an embellished version, as part of his atlas of the known world. Imperfect though these sources may be, they do give tantalising glimpses of the Scottish landscape and an indication not only of where there were woods, but sometimes what they were like.

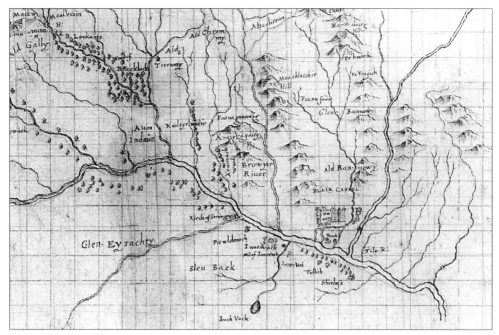

PLATE 4.1 *Timothy Pont's map of parts of Glen Garry, c. 1590. The strath appears only lightly wooded, even around Blair Castle, and the Falls of Bruar ('Browyhir River') are as bare of trees as in Burns's day two centuries later. There was, however, a substantial 'Wood of Blacklach' on the Allt Glas Choire where there is nothing now but open moor. (Reproduced by permission of the Trustees of the National Library of Scotland.)*

In the Lowlands, Pont leaves us with an impression of few woods of any size, but they often appear to be enclosed, which suggests they were valued and managed.

If the union of the crowns helped curb the lawlessness of the Border reivers, it would be another 150 years before the Highlands were completely peaceful. Here, the clan system, if not as powerful as it was in the days of the Lordship of the Isles, nevertheless retained a firm grip on society and its economy. This Celtic system, based on feasting and feuding, had the clan chief as territorial supremo boasting a kinship link (real or imaginary) with his tenants, rather than a Lowland landowner whose relationship with his dependants was based only on cash. The Highlands had a language and culture of its own. This, combined with inherent remoteness, gave the impression of a wild and dangerous region, isolated from the rest of Scotland, inhabited by violent skirted barbarians and economically underdeveloped. At least in the eyes of southerners!

Imagine, therefore, the young Timothy Pont, brought up in the relative comfort of a prominent churchman's farm near Culross in Fife, embarking on a fairly perilous journey into potentially hostile country, into a land characterised by huge expanses of rugged terrain, and a population, suspicious of 'sassenachs', speaking another language – he might as well have headed for the New World. His legacy is an extraordinary, if sometimes difficult to decipher, series of around forty manuscript maps, showing coastline, lochs and rivers, mountains, castles, settlements, woods and copious placenames.

The sketches show the Highland landscape, not covered in dense and continuous

PLATE 4.2 *Roy's military survey, c. 1750, of the Great Glen immediately above Fort William and north of Ben Nevis. Woodland is already scattered, with far less in the landscape than at the present day, when much of the southern portion of the map below the military road is under forestry plantation. On the other hand, most of the original woodland is still there, although the wood to the north (in Glen Loy) is partly replaced by non-native conifers. (British Library. Copyright reserved.)*

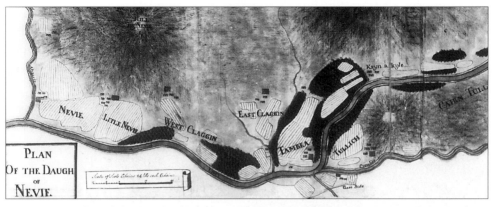

PLATE 4.3 *Plan of the Daugh of Nevie in Glen Livet, mid-eighteenth century. The riverside woods have been partly cleared for cultivation. (National Archives of Scotland RHP 2487.) (Reproduced by permission of the National Archives of Scotland.)*

forest, but with a similar, though more extensive, distribution to what we have today. Heading north, Pont might have passed along the familiar wooded glens and slopes at the foot of the Ochils. Onwards to Perth, and the ancient Methven wood overlooking the River Almond with the fertile and probably largely open Strathearn, to the west. Not until he reached the gateway to the Highlands at Dunkeld would he start to get the sense of lofty mountains and increasing woodland cover with oak-, ash- and birchwoods draping the valley sides of the Tay. He might then have chosen to travel west to Loch Tay, where he sketched the lochside as a mosaic of woods, settlements and castles, giving the impression not dissimilar in extent to what exists today, though many of the woods are now commercial conifer plantations. Here he was also taken by the 'fair salmonds, trouts, eeles, pearle', noting this on his sketch map.

Had he instead headed north, travelling through the heavily wooded Pass of Killiecrankie, he would surely have been impressed by the deep gorge and the crags covered in the full array of broadleaved trees. The Pass was not entirely wooded as it had enough open ground a hundred years later to permit a Highlander charge at the battle of Killiecrankie. After Blair Castle, its grounds already adorned with planted trees, the woods may have gradually petered out as Pont headed across the barrens of Drumochter, a wild but not unused land, dotted with sheilings, then on into Badenoch, Strathspey and the Caledonian pine forest. Perhaps this was first time he had ever walked in deep forest along shady paths, the air around him infused with the aroma of crushed pine needles. Interestingly, though, his map of this area is littered with placenames more than with tree symbols – people and woods in intimate proximity. Yet the forests were more extensive than today. At the beginning of the seventeenth century, travellers recorded a 'firr forest, 24 miles in compass' in the parish of Abernethy – rather bigger than today, and probably extending over the Pass of Ryvoan to touch Glenmore and Rothiemurchus, yet interspersed with houses and fields.

It's difficult to imagine what it was really like for Pont on these extensive travels. Most certainly a journey of discovery! What would have surprised him might not be the same as would catch our modern-day eye. He might have expected the Highlands

PLATE 4.4 *An old farmhouse at Kinchurdy, Boat of Garten, showing pine deal walls and cruck construction (and a corrugated iron roof). In Speyside, lavish use of wood was easy. (The Highland Council, Highland Folk Museum. Copyright reserved.)*

to be replete with forest – to find 'wildwood'. Was he as enchanted as we are today? Certainly he was keenly interested in nature. He often jotted down the abundance of the natural resources, like the fish of Loch Tay. He also mentions, game and wildfowl, wolves, and, not surprisingly, biting flies. In Sutherland, he noted 'heir yrons made'; mapping not only the location of this early ironworks, but also the place nearby where the iron ore was found.

Interestingly, the Scottish Parliament under James VI, only a few years later in 1609, prohibited new iron-making by Act of Parliament, having recently discovered (perhaps thanks to Pont) that there were 'either unknown or at least unprofitable and unused' woods in the Highlands and fearing they would be lost to unregulated exploitation. A couple of years later James nevertheless granted a licence for iron-working in the Highlands, to a Scot, Sir George Hay. Intriguingly, this ironworks site was Loch Maree, which had been discovered by Pont to be 'compasd about with many fair and tall woods as any in all the west of Scotland, in sum parts with hollyne in sum parts with fair and beautiful fyrrs of 60, 70, 80 foot of good and serviceable timber'. And, not just Scots pine and holly, but also oak, birch, ash, elm and aspen.

There are other early seventeenth-century descriptions of woods in the Highlands, which give the impression of significantly more diversity than we find in the same woods today. The general ecological distributions are little different, with birch characteristic of the north, oak in the west and Perthshire, with concentrations of Scots pine in the eastern Highlands, but also around the Great Glen, Ross-shire and North Perthshire. The difference was the great assortment of trees in any given area. Now we might stand on the shores of Loch an Eilean and feel close to the primeval forest, but, magnificent as they are, too much has happened to these woods for us to truly experience wildwood. Rothiemurchus and the other Strathspey woods were much more diverse than today. It is the early seventeenth-century woods like those of Loch Maree, which were full of holly, rowan, aspen – trees rarely seen in profusion today – that must have been closer to wildwood, but they too are now a shadow of their seventeenth-century glory.

Although it is true to say that most of the Highlands had lost its tree cover by 1600, nevertheless there were more and bigger woods, some of which have not survived to the present. Both the Lowlands and Highlands lost woods between 1600 and 1700, with more going thereafter. In the Lowlands, the Torwood, near Stirling, had been greatly reduced and Falkland wood was completely lost by 1700, claimed for cultivation after the best trees had been taken by Cromwell's army to build Perth Citadel in 1653. It is probable that Scots pine has diminished in extent, and not just in core areas. It is entirely possible that pine extended further north and west, possibly as far as Assynt. Elsewhere, the Pont map of Loch Leven and Glencoe describes the south shore of the Loch with 'many Firr Woods heir along' – long since disappeared.

What happened to Falkland wood gives us one possible explanation why Scotland's woodland cover continued to decline after 1600. How they were viewed, used and abused therefore must form the next part of the story of Scotland's woodland history.

The Use of Wood in Everyday Life

Glenmoriestoune is a verie profitable and fertill little glen . . . plenteous of corne and abundance of butter, cheese and milk and great long woods of firr trees doth grow in that countrey. And the river doeth transport big Jests and Cutts of timber to the fresh water Loghnes . . . and thereis ane little parish Church of timber in this countrey.

So said an early seventeenth-century visitor to Glen Moriston. This snapshot is illuminating. It confirms the presence of a Caledonian pinewood, which was being exploited and the timber floated down to Loch Ness, probably then on to Inverness. Furthermore, timber was being used in local building construction. There also is a remarkable sense of rural idyll, the abundance of farm produce not quite squaring with the wild, untamed and underdeveloped land described by nineteenth-century romantic writers. Instead we glimpse the settled, if not untroubled, nature of life during the first half of the seventeenth century. Rough times were ahead for Glen Moriston folk, but this reveals a bountiful agriculture with woods capable of providing for both local and external demands, without appearing to denude the resource. Certainly, a hundred years later, a map produced to assist with timber exploitation still noted pine-, birch- and oakwoods in the Glen, and there are still some pinewoods

PLATE 4.5 *A large Highland house in the process of being reconstructed at the Highland Folk Museum, Newtonmore. This is a 'caber house', with internal roof and walls made of poles. (The Highland Council, Highland Folk Museum. Copyright reserved.)*

in Glen Moriston today.

In certain respects this scene could be anywhere in Scotland, not restricted to the Highlands, because, to a greater or lesser extent, growing crops, rearing livestock and the need for wood products were universal elements of the Scottish rural way of life at that time. The relative importance of each activity depended on geography. Broadly speaking, fertile Lothian and Fife were engaged in arable farming to feed a burgeoning urban population, while the Borders and the Highlands depended on livestock, sheep in the former, cattle in the latter (in the seventeenth century at least). Everyone needed wood, but found it from very different sources. The Northern Isles, with no woods to speak of, used driftwood, and their buildings had more stone and turf in their construction. Most Lowland areas, as we have seen, had enough woods at least to meet farm needs. Depending on location and wealth, building needs might be otherwise supplied by imported timber. Even within the Highlands, there were regional variations. We have already heard of whole buildings of timber, usually where there were plentiful woods, as in Glen Moriston, whereas other areas used a combination of stone, turf and wood.

We live in an age of plastic, steel and concrete. Most folk have little idea how their house was constructed. Our needs are met by big DIY stores and a plethora of large building firms. We are disconnected from the natural resources that serve our basic needs, like heat, light, clothes, food, drink and a roof over our heads. This is a relatively modern phenomenon. The necessities of life, which come to us so easily today,

Plate 4.6 *Children outside their home at Letterfearn, Lochalsh, late in the nineteenth century. Notice the wooden barrels and wash tub as well as the wattle construction ('stake and rice') of the wall behind them. (Bob Charnley Collection.)*

were hard won by our ancestors. They had to source wood, cut it and make all their own farm and household equipment. Trees and what grew on and around them (from bough to bark, branch to berry, and lichens to sap) all had a use.

Even in the seventeenth century, timber was still a major constituent of urban houses. The houses of Edinburgh's old town were noted for being tall (six, seven, even an incredible fourteen storeys), and they were also characterised by overhanging timber facing, and interiors divided and lined by 'deals' (boards), usually pine. Many other Scottish towns had similar architecture (if not quite so tall). Dundee, Aberdeen, Inverness, Cupar, Kelso, even Glasgow were all built using considerable amounts of timber. This was much to the consternation of burgh councils, who feared catastrophe. Fire did break out in several of these towns, and Edinburgh at least could boast one fire in 1700, which was said by some to have been fiercer than the Great Fire of London. New building regulations laid down that frontages should be of stone, but interiors still needed enormous quantities of wood for rafters, joists and flooring. When the Georgian New Town of Edinburgh was built, it caused a surge in the timber trade.

Except perhaps for Inverness and some other northern towns, the timber for townhouse construction and for more substantial country houses and castles came from Scandinavia and the Baltic, later from North America. Adam Smith said that there was not a stick of Scottish wood in the New Town. Scotland's woodland resource simply could not supply the level of demand. Not very much has changed really: Britain remains dependent on timber imports. Then, as now, imports were cheaper and even where local supplies of pine, which was then the favoured tree for building, could be more readily secured, it was invariably regarded as lower quality.

In outlying rural areas, and the Highlands in particular, native trees were indispensable. A use was found for every part of all trees and shrubs, large and small, – in house building, for farm implements like ploughs and harrows, also household items like bowls, barrels, baskets, tubs and rudimentary furniture. Dyes could be procured from different parts of a tree, with alder alone yielding five colours. Bark was important for home tanning of leather, with travelling shoemakers coming round homes making up shoes and possibly other leather goods, though not as you might think saddles and harnesses, for they were invariably made from wood, birch twigs being twisted round to make reins and ropes. Indeed birch was 'the universal wood of the Scottish Highlanders' as one commentator remarked. Birch was, however, often the most abundant tree in those woods (also including alder, willow, hazel and rowan) that were traditionally accessible to local folk, so perhaps they had little choice.

Other favoured trees included ash, often referred to as the 'husbandman's tree' because of its suitability for farm tools such as ploughs and the *cas-chrom* (Highland spade). It was also used for shinty sticks. As well as being a source for dyes, alder was used in roofing. Hazel and willow were invaluable in providing the smaller flexible 'withies' or 'wands', which were woven into hurdles and baskets. Fences were not like our neat square post and wire fences today, but more usually what was termed 'stake and rice' – more like wattling, where twigs were woven round vertical stakes.

Oak was indisputably the most prized of all trees, but often reserved for the chief or laird, although the factor might allow it to be used for building mills, manses and boats, the crooked 'knees' of oak being particularly valued for the last named. In areas where oak was particularly common like Argyll or Perthshire, some of the smaller trees and branches might have found their way into poor tenants' homes. As oak bark grew in value during the eighteenth century, the small timber of stripped trees may well have been offered or sold to tenants to encourage house improvements. However, when this small oak timber could be more profitably sold as charcoal feeding the Highland ironworks, this too became off limits. There were many, no doubt, who found such conditions too restrictive, and helped themselves. Three things in life were free, said a Highland proverb, a fish in the stream, a deer on the hill and a tree in the wood.

In rural areas, it was not only house and farm that were served by woods. Today, as we zip around the countryside in our cars, it is hard to imagine how long and difficult even simple journeys, say Perth to Inverness, would have been 400 years ago, when Pont was travelling. We travel on the stilted A9 over the Pass of Killiecrankie in a few minutes, whereas Pont might have taken an hour on horseback negotiating burns and boulders, along a rough track.

Wheeled transport had limited use in the old Highland economy. Moving goods, like barrels of whisky or salmon, was normally by packhorse or perhaps tumbler – a primitive two-wheeled cart with solid wheels. The wheel-less Highland 'slype' (sledge), framed by two sapling trunks, was used on rough ground, often to carry peat down from the hills. All were made of wood.

By 1769, when Thomas Pennant toured the Highlands, roads were beginning to improve. He seemed impressed, for instance, on travelling up Loch Tay, with the '32 bridges built by Lord Breadalbane, on the north side of the Loch alone'. This was obviously a recent improvement: some bridges may have simply involved rough boughs and branches, but the larger ones needed substantial timber, some being entirely constructed of wood, and even stone bridges needed scaffolding. Oak was favoured for the bigger bridges, where strength was critical, as it was also for boatbuilding.

At the start of the seventeenth century the age of the Highland war galley or birlinn was in ebb, but such symbols of prestige were still being built. In 1613, 'ane gaillay of twentie foure airis with her sailing and rowing gear gud and sufficient within ye spaice of ane yeir' formed part of a marriage agreement between two clan chiefs. Boatbuilding was not just important on the coasts, it was also very much a key activity on inland lochs. With the difficulties of overland transport of goods, boats were a much used alternative. Even as late as 1799 (when roads were much improved), it is recorded that a boatwright on Lochtayside was given permission to get oak from nearby woods for a new boat.

Interestingly, the farm where this Lochtayside boatwright procured his oak was called Camuschurich, meaning bay of the currach or coracle. Currachs were the other end of the boatcraft spectrum from war galleys. They were essentially hide-covered baskets and ranged from seagoing 'boat-shaped' vessels, often linked to the times of Columba, to the much smaller round or oval ones. The latter were still in use up to the nineteenth century on rivers, especially the Spey, where one of their uses was to guide pine timber being floated down river. What prompted the Lochtayside inhabitants to name Camuschurich as they did is lost in the mists of time, but it is not inconceivable that the extensive and diverse wood that reaches down to the lochshore might have had something to do with it.

Placenames can reveal much about past associations, giving insight beyond the physical character of a place. The difficulty is knowing at what point they were created and why. If today, there were no trees near Camuschurich, we might have more to puzzle over. Another example from Loch Tay of the revealing nature of placenames is Croft-na-caber or croft of the rafters. Now a watersports centre, the woods around it, dominated by birch and hazel, may have long been known as the place to get your house rafters. And although we associate cabers with those we see being tossed at Highland Games, in past house construction, they were in fact much smaller than the doughty trunks of the tossing variety. Some may also know Croft-na-caber as the location for Scotland's only replica crannog, an earlier 'house' style, which also needed many 'cabers' – an interesting twist of fate!

When we think of Highland vernacular buildings, we often assume that 'black houses' were the universal style, but there was in fact much variation. We only have to look back to seventeenth-century Glen Moriston for evidence. If a church could be made entirely of timber, then so too could ordinary houses, especially where there was an abundance of wood. One Strathspey laird's woodland adviser commented in

PLATE 4.7 *The interior of a Highland house from an engraving by A. Duncan, mid-nineteenth century. The family has a number of possessions, almost all of them wooden: furniture, barrel, churn, tub, spinning-wheel, candlestick and walking stick, perhaps the plate and cups as well. (The Highland Council, Highland Folk Museum. Copyright reserved.)*

1762 'there is as much Wood destroyd in building Walls of houses, as might serve a whole Nation. Stone walls would do much better.' Here, as in Glen Moriston, whole houses were constructed of timber, probably a combination of pine and birch. Also, with the possibility of getting sawn pine, a mix of 'deal' and 'creel' might have been used. That is, use of boards and also wattle, with perhaps only rough stone foundations. Deeside houses were probably similar, where local trading in 'deals' by farm tenants, particularly with Perth and Angus, suggests that sawn wood again was more common and might be used by local folk in their house construction.

Further north and west, where sawmills were less common, only important buildings like those of the chiefs or his tacksmen used 'deals', like Cameron of Lochiel, whose house in 1723 was described as 'all built with fir-planks, the handsomest of that kind in Britain', until of course the redcoats rased it in 1747. Otherwise, houses in the wooded parts of the Highlands are likely to have been 'creel houses', where poles were set into the ground and flexible saplings woven between these uprights, up to about eight feet in height. Gables would be similar, and houses may have been more oval than rectangular. The most substantial timber would be the couples (crucks), which were pinned at the apex of the roof at either end of the house and then connected and strengthened by horizontal purlins (beams), infilled by cabers and topped by thatch of heather, bracken or turf. The wattle walls might then be 'insulated' using turf,

PLATE 4.8 *Three types of Highland cart as seen by Edward Burt in the central Highlands, c. 1720. Notice the solid wooden wheels and the great load of sticks, perhaps for making wattle. (St Andrews University Library.)*

although even in the late 1800s, bare 'creel' walls could still be seen. Different parts of the Highlands would use turf and wattle for walls in different proportions, depending on local availability of the raw materials.

Not only were Highland houses of farmers and crofters rarely made of stone in the seventeenth and eighteenth centuries, but they also continued to be a transient feature of the landscape. The population was not static; the many uncertainties of life did not encourage permanency. It was only when the 'Improvement' movement started to permeate the Highlands in the late eighteenth century that lairds and their advisers pushed the use of stone. Lairds and chiefs, or perhaps we should now call them landlords and proprietors (for that is what they became during this major socio-economic transition), tried various methods to persuade their tenants to build with stone. In Lochaber, the Duke of Gordon was advised in 1767, 'if the tenants were obliged by their tacks to build stone houses would preserve the wood & be much for their advantage and profit'. A similar argument was made in Strathspey, where it was reasoned that the tenants 'knowing they can get Timber whenever they call, neglect both their Walls and Roofs: they throw down the one for muck [manure] and the other for fire . . . costs them a great deal of time that might be better employ'd and ruins the Lairds Meadow-Ground and Woods – whereas if they were obliged to buy their Wood, they would take better care of it'. 'Profit' was becoming a key word as market forces began to cast aside traditional Highland values.

In the Lowlands, farmers were long used to buying wood and other necessities. Many of the private contracts and public roups (auctions) of woods were selling to local farmers (perhaps in loose partnerships) or tradesmen, who might then process the raw material and sell a product with a profit. Woods had therefore been managed to produce a commodity for much longer. Highland landowners, as they became assimilated into Lowland, indeed English, society, introduced such sale methods on their Highland estates. So, when woods started to develop a greater economic value and chiefs and lairds sought profit to support their increasingly lavish lifestyles, access to the woods for local use became more problematical. Encouraging the use of stone, as we have seen, reduced one pressure on the woods. It also enabled them to be turned over to more profitable operations, and subjected them to new pressures.

Eventually, Highland farmers had to pay for wood and by the late eighteenth century, new, neat stone built villages were popping up, like Inveraray and Grantown, housing tradesmen and estate workers, making and selling their wares and services. The farms also changed. Hitherto, clusters of houses, where several families jointly farmed the surrounding land, with access to common hill grazing, had become by the 1850s, either single tenancy small holdings or large sheep farms. These new buildings still needed roof timbers and the farmers needed their tools, but how they procured them changed. With the development of plantation forestry, needs could be met cheaply by sawmill waste and thinnings or by imported timber now carried cheaply overland on the railway; the tools were increasingly made in factories and sold to the countryside. Gradually the emphasis shifted away from almost total reliance on the 'natural' woods. The 'Improvers' won through and rural activities started to echo those of the Lowlands, although not without considerable human suffering.

Gradually, perhaps almost imperceptibly, those who lived off the land became divorced from woodcraft.

Woodland and Farming

As long as the woods only had to support local timber and other woodland produce needs, even with the limited local trade that existed, they might well have continued unchanged in extent and composition. Recovery from felling was helped by the ability of broadleaved trees to grow again, and again, from stumps after cutting. And pine, which does not grow from stumps, can regenerate prolifically, given the right conditions of open, thinly vegetated ground. However, there were other demands on the semi-natural woods.

As demonstrated by the early seventeenth-century snapshot of Glen Moriston, crops and livestock were the mainstay of Scotland's rural economy. The inhabitants of the expanding urban centres relied on the rural hinterland to feed them, and for wool and hides to process, as industrialisation developed. Earlier chapters have shown the part farming played, since prehistoric times, in the deforestation of Scotland. Why

PLATE 4.9 *A Highland chair made from the knee of a tree. (The Highland Council, Highland Folk Museum. Copyright reserved.)*

PLATE 4.10 *A fisherman and his family at Nairn in the late nineteenth century. People were often surrounded by wooden materials at work, as well as at home. (Bob Charnley Collection.)*

Plate 4.11 *Traditional basket making was still being practised in the twentieth century. (National Museums of Scotland.)*

would it be any different after 1600?

By then, the remaining Lowland woods were generally on ground not best suited to agriculture, in gorges or wet riverside banks, like Roslin Glen or Ayr Gorge. Some that remained on flatter, drier ground, like Dalkeith or Cadzow, were the vestiges of hunting preserves, where trees provided shelter for deer and other quarry. By the early seventeenth century, when deer forests became less fashionable, some of these 'deer' pastures were incorporated into the ornamental parkland of the great houses, like Hamilton or Dalkeith. Other woodland remnants, less visible from the noble lord's balcony, may have been turned over to cattle and sheep pastures, while the temptation of trading big trees for ready cash and new arable land with no impediments to the plough dealt the final axe blow to the more unfortunate woods.

The decision-making process was complex and depended on the perceived priorities at the time. For example, when, in 1743, an owner near Tulliallan sold his wood (i.e. the trees, not the land) to a tradesman, there was a condition that he was to 'holl out at the root the whole trees and roots of trees . . . and to put the ground in such a condition as a plough can labour the same as other arable ground'. If you consider how big and tough roots can be, this was no mean feat. In this case, the local folk may also have been happy with the change, as they would have less requirement for local-grown wood than previously – with easy access to markets and sea ports, where substitutes could be procured. However, this would not always be the case, and there are many instances where the landowner took a decision about woodland that adversely affected the locals.

Local folk around Methven wood might have preferred some of it grubbed up like

Tulliallan, but in this case it continued to be valued and managed for coppice. When the value of oak coppice increased in the eighteenth century, we must assume that the value of such woods was deemed greater than it would be as agricultural land, either arable or pasture – a cost–benefit analysis. The fact that it was also close to a manufacturing town, Perth, would also have come into the decision-making process. Another Perthshire wood, Kincardine wood near Auchterarder, occupied land that could be converted to agricultural land, although not prime arable. But, it was also traditionally worked for oak coppice, and like Methven wood survived this period, although cultivated clearings were created within it. Like the Torwood, its extent had been reduced somewhat by 1750, but unlike the Torwood, it survived to the present day. It is probably significant that Kincardine was under the control of the Duke of Montrose, thought to be the largest owner of oak coppice in west–central Scotland in the mid-eighteenth century.

With the relentless demand for grain in a period of population growth, the pressure to convert woodland to cultivable land is quite understandable, especially in the highly fertile grain-growing areas of the Lowlands. The relationship between woodland and animal husbandry is less clear-cut. This is currently perhaps the most controversial aspect of Scotland's woodland history – the role and impact of grazing in woodlands. On the face of it, it is quite straightforward – animals eat trees and therefore are harmful to woodland. If only it was that simple!

As we have seen in the previous chapter, despite the presence of royal deer forests in the Borders, affording trees some protection, the value of large flocks of sheep was too great to prevent woodland decline in the fourteenth century. However, where woodland survives today, in a structure suggesting the marrying of woodland and livestock management (i.e. scattered trees with spreading crowns, possibly showing signs of past cutting of branches above the reach of animals), the question arises – was there clearly defined and regulated wood pasture? Or was it just that animals through necessity found shelter and grazing in woods? It may simply have been a never-ending battle to restrict animals from damaging woodland – a battle, which if we look at Highland woods today, was repeatedly lost.

The very fact that Scotland had so little woodland left in 1600 or in 1750, unlike most other European countries, and indeed England, suggests that agricultural production was paramount. The ability to pay rent depended on agricultural surpluses realising cash. In some parts of the Highlands, as late as the second half of the eighteenth century, rent was still paid in corn, cheese, cattle and sheep. Cattle became an increasingly important money-earner from the seventeenth century onwards, particularly in the Highlands and in Galloway.

Lowland woods by the seventeenth century seem to fall into one of three groups. There were those, like Methven or Pressmennan, traditionally managed for coppice and timber since medieval times; there were remnants, which had been part of hunting forests and continued as either wood pasture or parkland, such as Cadzow; and then there were those which, for various reasons, were still poorly managed and open, at least, to seasonal grazing. The future of this latter group was invariably sealed during this period. Some were lost to agriculture as happened to woods in Strathbogie, others became valuable and were enclosed like those of Eskdale, when an ironworks was set up at Canonbie in the opening decades of the eighteenth century.

Myreton wood, near Alva, might fall into another category – continuing to be open to grazing, but surviving. The old ash at Myreton today show signs of past pollarding (being cut above the reach of animals). If we define wood pasture as a land use where trees and pasture are managed in order to perpetuate both indefinitely, then Myreton may be one surviving example. Cadzow and Dalkeith parks are clearly also examples, but there is more clear-cut evidence for them. It is the history of the unenclosed upland woods, like Myreton or Glen Finglas (see Plate 4.12) in the Trossachs, that remains obscure.

The documentary evidence often focuses on problems of livestock incursions and damage to trees, rather than how the likes of Myreton came about. Invariably this evidence is an estate's record of events, and as such, unfortunately, we rarely hear the voice of ordinary folk, except when involved in the bargaining process relating to rent, or in explaining their misdemeanours. Since it was the tenant, cottar and herd, who made the day-to-day decisions about how a wood was used, historians are left to puzzle over wood-pasture origins and development, particularly in upland areas.

Highlanders were pastoral farmers who could not survive without their livestock, primarily sheep, cattle, goats and horses. To the Highlander, woods were particularly important as a critical source of winter shelter and spring grazing, where cattle could get their first bite of new grass before they were sent to the hill sheilings. As long as the animals did not do too much damage to seedlings and saplings, which would later replace the older trees needed for all the essentials of life, the woods would continue to flourish. But a tree's ability to put out new shoots after cutting would be negated if hungry mouths then sheared them back. And, if this happened, year after year, eventually there would be no young to replace old. Woods would become pasture with scattered trees, old and decaying, then lost forever.

As with the Lowlands, there seems to have been three categories of woodland. There was a small proportion, mainly oakwoods, which were enclosed and managed, at least since the sixteenth century. We know that some of the oakwoods on Lochtayside fall into this category.

A second group involved hill woods, often associated with royal and baronial hunting forests like the Forests of Mar, Mamlorne and Atholl. Their owners might at times try to limit grazing by tenants' stock, but they were seldom successful in doing so for long. In practical terms, they merged with a third group – unenclosed and traditionally accessible to stock, particularly in winter. This group also included most of the Caledonian pinewoods, which were grazed and exploited for timber, without fencing, right up to the late eighteenth century.

Concessions were made to accommodate the needs of livestock even in woods which had been fenced off. For example, cattle or sheep might be allowed back into the wood, five or six years after it was cut and enclosed. There might also be regulated access at particular times of the year for specific animals (e.g. calves or ewes).

As more and more woods came into the category of 'enclosed' during the eigh-teenth and early nineteenth centuries, the pressure intensified on 'unenclosed' wood

PLATE 4.12 *Ancient wood pasture on Woodland Trust land at Glen Finglas. The trees, hundreds of years old, appear to have been pollarded just out of reach of the animals. Glen Finglas was once a royal hunting forest. (Peter Quelch.)*

and tree-less pastures. The value of semi-natural woodland increased from the demand for bark to tan leather, particularly during the Napoleonic Wars, just as the same demand for leather (and beef) was driving the expansion of the black cattle trade, and putting greater pressure on traditional pasture.

Hill woods and pasture of all kinds were therefore increasingly affected by the growing numbers of black cattle through the seventeenth and eighteenth centuries, but they shared the uplands with other animals. We have seen how 'royal' hunting forests became redundant in the Lowlands after the union of crowns, but Highland chiefs and lairds continued to enjoy such pursuits. Some of these hunting forests continued to be important refuges for deer, like the Forest of Mamlorne in the hills between Glenlochay and Tyndrum.

One of Gaeldom's most important nature poets, Duncan Ban MacIntyre, was, in fact, employed as a forester (gamekeeper) in Mamlorne by the Earl of Breadalbane until the 1760s. We know from his poetry in praise of his native hills that hunting deer was still a lordly pastime. We also get a vivid picture of the forest supporting lush pastures and fruitful woods of 'stalky bushes, full of boughs and sprays; with verdant saplings, shoots growing densely, and foliage shrouding the head of the trunks'. Within twenty years, he was lamenting the passing of the hunting tradition, and especially the degradation of the hills; his favourite corrie had been 'blighted', 'cropped to the ground' and the deer were gone. The hill woods had suffered too, 'neither wood nor heath endured there'. He was in this case decrying the effects of the coming of the sheep, the Lowland black face and cheviot breeds that became the scourge of the Highlands. Then, as well as cattle, deer and sheep, the hills supported increasing numbers of horses and goats, the latter a particular danger to the regrowth. The lairds' factors tried, not always successfully, to discourage goats, but they were a useful subsistence animal for poor tenants (see Plate 4.13).

Some hunting forests, like those of the upper Spey and Dee catchments, still supported extensive pine- and birchwoods. A pasturing tradition seems always to have existed in these forests. Attempts by proprietors to curb grazing in them were not so much linked to deteriorating woods, but developed as the possibilities of greater financial return from the pine timber became a possibility. Again, this increasing timber exploitation came as the population was increasing, so that areas within some of these forests, particularly the forests of Mar and Atholl, which had supported sheilings, at least on their margins, were in the early decades of the eighteenth century becoming permanently settled and even cultivated.

One famous dispute involved the Forest of Mar, which until the first Jacobite rebellion of 1715, was held by the Earl of Mar. A famous medieval deer-hunting venue, the Earls had feued out to neighbouring lairds and tenants the use of much of the land and also timber for local building. The new owner, the Earl of Fife, recognising the value of his pinewoods, and concerned about the effects of grazing, cultivation and local timber use, embarked on a lengthy legal battle to regain control. One of his concerns was that because 'these Highland fir woods gradually shift their stances', cultivation at their margins would prevent them naturally regenerating beyond the woodland edge. Interestingly, part of the deal involved the Earl giving up his interest in Ballochbuie pinewood, in return for the recipient, Farquharson of Invercauld, relinquishing his feu rights in the other Mar woods. Having gained control of the whole pinewood of Ballochbuie, the Farquharsons, late in the nineteenth century, agreed to sell the wood to an Aberdeen timber merchant. This might have had devastating consequences, had it not been for Queen Victoria stepping in and buying the picturesque ancient wood herself.

For all the legal protection which was available to hunting forest proprietors to try to stave off the worst deprivations of livestock, the grazing imperative usually defeated them. Increased stocking of cattle, combined with the continuing presence of goats and horses, must have played a part in reducing their tree cover and ultimately the grazing value.

Beyond the partial protection of hunting forests, there was a third group – the unenclosed woods of straths, glens and lochshore. These were the woods that bore the pressure of winter grazing. Livestock rearing methods varied throughout the Highlands, depending on climate and available resources for sustaining the animals. In the warmer west coast, animals might spend the winter on stubble fields or in woods, while harsher winters in the east and north would encourage 'sharing' of the house with the animals. Animals were not just left to get on with it either; unlike today, herding was commonplace, usually the job of the younger family members. Country folk, steered by their lairds and tradition, had their own way of managing stock and woods to produce grass and timber. These ways, largely unknown to us,

PLATE 4.13 *Tending animals near Glen Finglas, from the decoration on an estate map, 1782. Close herding would restrict the damage that grazing could do to a wood, though the tree on the right is not very healthy. Note the goat in the centre. (National Archives of Scotland, RHP 6001.) (Reproduced by permission of the National Archives of Scotland.)*

might have sufficed to keep the woods in good heart, had it not been for forces at work beyond their ken, which changed their way of life and their woods.

The gradual deterioration of woods and pasture, perhaps unseen by those living among them, day in day out, did eventually begin to show. The Sutherland straths in the seventeenth century were noted for their 'woods, grass, corns, cattell and deer, both pleasant and profitable', including Strathnaver, which was 'weel stoored with wood'. Within a hundred years, largely unaffected by enclosure and management for commercial purposes, these wood pastures were clearly suffering. Not only were the woods deteriorating, but as the Kildonan minister commented in 1812, there was 'a degeneracy of black cattle in the parts that were formerly covered with wood'. The coming of the sheep to these straths was just the final straw that broke the people and the woods of this northern county.

Further south, Duncan Ban MacIntyre was seeing in the Glenlochay hills of the 1780s, the effects of an earlier, if less brutal, change in the fortunes of woods and pasture. Sheep were finishing the job begun, perhaps a century earlier, by goats, horses, cattle and external forces beyond anyone's control. However, down the hill and beyond this bard's precious hills, in Argyll and Perthshire, these same forces were instrumental in persuading landowners to protect their thicker, compact and more oak-dominated woods. Entrepreneurs had arrived from the Lowlands, England and Ireland, to tempt them with moneymaking opportunities, which would help with the building of their new stately piles. Some woods may have stayed open to meet the needs of country folk, but many were being fenced. For example, of the 2,030 acres of woodland in Argyll owned by the Earl of Breadalbane in 1786, only 150 acres were unenclosed. As we shall see later on, these developments, on the face of it, beneficial for the woods, would only squeeze the people more, and conflict was inevitable.

CHAPTER FIVE

Using the Woods, 1600–1850
(2) Managing for Profit

MAIRI STEWART

INTRODUCTION

Some might argue that the Napoleonic Wars, between the 1793 and 1815, were a boon for native woodlands. External imports of wood products were disrupted; inflation was high but so were prices. Coppice woods were in demand and with that came more careful attention to ensure that the commodity was sustained. The more accessible pinewoods had long been sold for profit, but now, cut off from the Baltic and Norway, they seemed more attractive than ever for purchasers.

COPPICE WOODS

Trade in coppice produce had long existed in the Lowlands and many woods were actively managed to yield bark and timber. Twenty-four years was thought the ideal cutting age for good-quality bark. Sale records indicate that buying and selling coppice produce was also commonplace in the seventeenth century along the southern fringes of the Highlands, where closeness to market towns such as Forfar, Perth, Kinross, Alloa and Stirling, helped to overcome transport costs. As the economy developed, not only in Scotland but also in England and Ireland, so the scale of transactions increased and a wider area of the Highlands became involved.

Bark was by far the most lucrative coppice product, not least because it was light and easily transportable. The trade in the early years of the seventeenth century was not huge, and most leather was for local consumption. Early records of wood sales confirm that in the Lowlands woods were being sold to a range of merchants, farmers and tradesmen, including millers, maltmen, bakers and wrights. The sales usually entailed 'the hail berk and beuche [boughs]' of the wood. Bark was therefore always part of the sale, and occasionally we see shoemakers or tanners purchasing directly (often in partnership with others). However, local budding entrepreneurs were probably the main purchasers, then selling on the different parts of the trees to specialist craftsmen.

By 1700, the tanning trade had become more organised and skilled, concentrating on heavy leather (for harnesses, shoes, belts etc.) from cattle hide, with oak bark as the favoured tanning agent, and Irish buyers were appearing along the coast from the Solway to Fort William, seeking bark for their own burgeoning tanning trade. By far the biggest internal market for bark and leather was Edinburgh, then the principal town of Scotland. Woods close to Edinburgh, like Humbie and Roslin, became particularly valuable, and their owners must have been rubbing their hands. Selling the timber would also be easier for them with fewer miles to cover than those from outlying districts. That said, the other Scottish burgh towns also supported tanneries and associated trades: evidence for this can still be seen today in street names, like the Skinnergate in Perth.

Pressmennan wood in East Lothian was traditionally coppiced, probably from medieval times. From sale contracts, we can get an idea of the relative values of the different trees. In 1797, for example, oak timber was sold standing at 1s 4d, while ash sold for 1s 6d, which illustrates the value of ash, but it was the additional sale of oak bark that puts into perspective the value of oak. Over 5,000 oak trees were in that sale, making almost £900, whereas the owner could only muster 162 ash trees. This was at a time when the price for bark was starting to soar, trebling between 1790 and 1813. It is often said that war creates prosperity. Falling back on Scotland's natural resources to aid the war effort certainly raised the value of woods. This in turn heightened the awareness of owners and advisers of the need to look after this, now, precious resource, if they were to continue to profit from them. The effect on woodland management was undeniable.

PLATE 5.1 *Coppiced oak woodland, with standards, in Worcestershire. This technique was generally used in eighteenth-century Scotland to produce tanbark, charcoal and timber. (Peter Quelch.)*

PLATE 5.2 *The Wood of Cree on the Solway, the largest ancient oakwood in the Lowlands, but typical of those used sustainably for centuries. The clearing is the site of an old charcoal hearth. (Peter Quelch.)*

Woodland scarcity and proximity to markets had long encouraged coppice management in the Lowlands. Management usually involved cutting all or part of the woodland (depending on size) twenty to twenty-five years after the previous cutting. Only a small proportion of trees (reserves) would be allowed to grow to the size that we today are used to seeing around us. Most contracts tried to regulate the care necessary during cutting, to avoid stumps rotting or young growth being damaged. Apart from large woods, contracts usually stipulated two to ten years of annual felling operations – most woods were simply not big enough to support economic annual cutting over the rotation period. Down payments would be expected, followed by annual payments – it was not unknown for buyers to go out of business before the contract was up. The buyer would often get grazing for their horses (particularly important in Highland contracts where feed was scarce). The seller would often expect brushwood to be cut for building the 'stake and rice' fence around the recently cut 'haggs' (section of wood). More often than not, the seller had to bear the expense of fencing and also to provide transport (using farm labour) to take the bark, but not the timber, to the tannery or other manufacturing destination.

Management of coppice in the Highlands was usually more slapdash than in the Lowlands, with fewer cutting regulations and enclosure less often demanded. As market demand reached its peak, however, many more woods in the Highlands came under coppice management using Lowland methods, if still a little more haphazardly applied than in Lowland woods. The difficulty lay in finding a market for the 'peeled' trees, which would create enough profit to encourage the landowner to enclose, perhaps for the first time. Not only that, Highland lairds still had the added complication of accommodating their tenants winter pasture and timber needs. The trend towards increasing commercial coppice management was undoubtedly based on the profit motive. However, many things that happened to woods must simply be put down to accident and not necessarily design. This was influenced no doubt by the owner's individual character – if he was by nature tight-fisted or generous, weak-willed or tyrannical, a lover of nature or driven by greed.

Nevertheless, the successful exploitation of Highland woods remained dependent on location, particularly for realising profit from coppice timber. Bulk overland travel was notoriously difficult in the Highlands. Therefore the areas that were particularly successful were closer to timber markets, including south-west Perthshire, Stirlingshire, Dunbartonshire and coastal Argyll. Industrial developments in the central belt created demand, particularly as the textile industry developed in the nineteenth century.

The focus so far has been on bark, without doubt a key commodity during this period, but a rather obscure product for modern-day minds to visualise. Scottish summers may not be getting any better, but come the school holidays every adult starts thinking optimistically of lazy sunny afternoons out on the patio with the barbecue sizzling the sausages. Barbecues we know about, charcoal we buy and use, although today much of it comes from far-off places, where highly diverse forests are destroyed to feed our demand. Charcoal, created by the process of 'cooking' wood slowly, has been used in Scotland for many centuries. It has played a key role in Highland woodland history, linked to the development of ironworks by the much-maligned English ironmasters and Irish adventurers. The charcoal story is certainly intriguing, the characters colourful and the woods, rather than always suffering at their hands, often came out of it, if not unscathed, generally intact.

The most famous of the Highland ironworks, and most successful, was the Lorne furnace at Bonawe on Loch Etiveside in Argyll, which opened for business around 1753 and continued blasting out iron until 1876 (see Fig. 5.1 and Plate 5.3). Today, a fascinating place to visit, its manicured grounds mask the dirt and grime of furious industrial activity supporting, in its heyday, 600 seasonal workers, mainly working in the woods stretching as far north as Morar and south to Knapdale and including woods on Jura and Mull. Around a dozen permanent employees worked at the furnace itself, many of them coming up from the south.

By understanding its success, we can see in hindsight, why most of the other Highland ironworks were relatively short-lived. The Lorne Furnace Company was set up by a Lancashire firm, which also ran charcoal blast furnaces at Furness in Lancashire. Its success, where others failed, was probably due to a combination of luck, being in the right place at the right time and able management, linked to a successful English operation. The luck came from persuading two local landowners, Campbell of Lochnell and the Earl of Breadalbane to agree to long contracts to supply coppice charcoal, the former for 110 years. Either Lochnell was short-sighted and feckless, or he was greedy

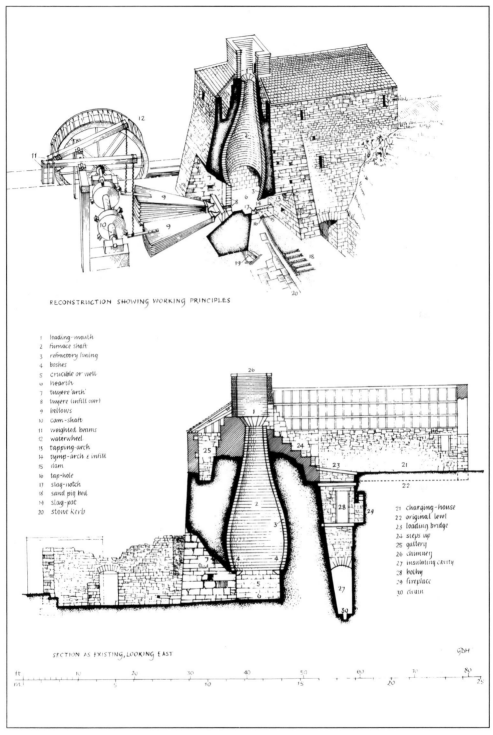

PLATE 5.3 *Bonawe Ironworks, Argyll, (above) the furnace and (opposite) the charcoal store. This was the most ambitious of the English ironworks in the Highlands, and operated from 1753 until 1876. Despite the immense size of the charcoal store, there was at least as much land under oakwood in Argyll when they left as when they came. (Royal Commission on the Ancient and Historical Monuments of Scotland: Crown copyright.)*

and desperate. Whichever, he must have rued the day he agreed to this contract, he saw, over the ensuing years, his neighbours agree short-term contracts which steadily gained in value.

The Bonawe managers appear to have run a fairly business-like operation, which cannot be said for some of the other Highland operations. One iron-making venture, set up by the ill-fated York Buildings Company at Abernethy in Strathspey in the late 1720s, lasted just over five years. Unusually, it was based on pine charcoal, and reputedly had to haul the iron ore over the Cairngorms, via the Lecht. They could not exactly float this commodity down the Spey. The York Buildings Company was involved in various other dubious ventures around this time and has been described as having neither wisdom in speculation, nor skill in management.

Speculation often seems to have been the name of the game, and lack of management know-how may have also contributed to the failure of an earlier furnace on Lochnell's lands at Glen Kinglas, further up Loch Etive from Bonawe. This one was set up by the Irish, Captain Arthur Galbraith, Roger Murphey (a tanner) and three other compatriots. Here at least, there was more chance of success than the Abernethy set-up, with reasonable access to the woodland resource and coastal transport. The

operation is thought to have lasted no more than fifteen years, and may have succumbed to a slump in trade in the late 1730s. The fact that Murphey was hanged for
murder in 1732, and Galbraith was complaining about his partners soon after, does
suggest that senior management was a little unreliable!

Another Irish initiative near Achray in the Trossachs, early in the eighteenth century,
also became unstuck, leaving a trail of unpaid wages. It may not have been a furnace,
rather a forge, but it still depended on the birchwoods of Loch Katrine and the
Buchanan and Menteith woods, under the Duke of Montrose's control. Two other
short-lived iron furnaces were set up: one on Cameron of Lochiel's land, at the end
of the seventeenth century, and another at Invergarry in the 1720s. The latter, again,
suffered the problem of being inland, making transport costs high.

Murphey and another Irish partner Edward Nixon, as well as having interests in the
Glenorchy pinewoods (little of which remain today), also set up the Letterewe
Company to exploit the oakwoods on the north shore of Loch Maree in the 1730s –
yet again another badly managed venture. This did not involve an iron furnace, but
over a hundred years earlier, Letterewe was the site of not one but two, possibly three,
blast furnaces. As already mentioned, Sir George Hay was given a licence to make iron
by James VI. He chose Letterewe. It is not clear how long this lasted, perhaps no more
than fifteen years or as much as fifty years, although intermittently.

There are several reasons why so many of these early ironworks did not last long.
Quite simply, the economic conditions were not yet right. It was not really until the
birth of the Industrial Revolution in the late eighteenth century that demand pushed

up prices and the English charcoal iron smelting industry needed more capacity. The Lorne Furnace Company, and another ironworks set up on Loch Fyneside (becoming known as Furnace) around the same time, came to the Highlands at the right time to catch this demand, the latter operating for sixty years. It is also worth bearing in mind that these earlier furnaces were 'planted' into areas unaccustomed to outsiders and potentially hostile, particularly if they were exploiting 'their woods' on a scale not seen before. These were still turbulent times, with Highlanders living in a world beyond the firm grip of central government and certainly not unfamiliar with violence and thieving. The chiefs, who were selling the woods, may have persuaded their own clansmen to leave the strangers alone, but that might not stop a rival neighbouring clan from picking off the 'sassenachs' and their goods as they made for the safety of towns and ports.

All of these ironworks consumed charcoal from coppice wood, insignificant compared to the 160 furnaces working in England between 1600 and 1800, but nevertheless they will have contributed to the changing nature of Highland woods. The woods of Loch Maree would never again be described as growing 'plentie of very fair firr, hollyn, oak, elme, ashe, birk and quaking asp, most high, even, thicke and great . . . [with] . . . great oakes, whaer may be sawin out planks of 4 sumtyms 5 foot broad'. The same might be said for the deciduous woods of Glengarry, Glen Moriston, Loch Eil, Glen Kinglas and Glenorchy. What effect did the ironmasters have on the woods? Some were haphazard in their approach, like Murphey and Company, others systematic, like the Lorne Furnace Company, who maintained for over a hundred years much of the woodland resource of Argyll and parts of Lochaber.

With a preferred rotation of twenty-four years, this would have meant at least four separate cuttings. It would certainly not be in their interests to ignore coppice management methods, which, after all, they would have been used to in England. The process depended on a steady fuel supply, which by and large was maintained over the years of operation. Today, some of the woods that supplied the Bonawe furnace may be in poor condition, but those woods that were directly managed under the Lochnell contracts, around Muckairn, remain remarkably intact, and support some of the best examples of Atlantic oak woodland.

Most of the woods that supplied the Bonawe furnace, however, were outside the company's direct control and depended on the owners' commitment to enclose after cutting. Standards of management were not universally high. There were some landowners, like the Duke of Argyll, who controlled a substantial area of woodland, who did make some attempt at enclosure. The Earl of Breadalbane, as we have seen, had almost all his Argyll woods enclosed by 1786, and other lesser Argyll lairds, like Campbell of Barcaldine and Sir James Riddell, the 'nouveau riche' owner of Sunart, were also enclosing their woods. The latter had added impetus, as he had a lead mine at Strontian also requiring charcoal. But how long would they remain enclosed, particularly after the ironmasters left, to be replaced by the flockmasters?

Some Highland areas were never really in a position to capitalise fully from the Napoleonic War boom. They were able to sell bark, but the timber remained for local use. With an ever-increasing population, barely subsisting on their black cattle, the incidence of enclosure was much less in such districts. This not only affected the more obvious isolated areas like the inland straths of Sutherland and Wester Ross, but surprisingly even the Earl of Breadalbane's Perthshire estate, centred on Lochtayside, only some forty miles from Perth. From around the 1780s, on Lochtayside, a regular system of cutting and enclosing woods was adopted, at least for those woods where oak was common. Bark was the most valuable product, but Lochtayside was just too far from the Bonawe ironworks or the Perth and Stirling timber and charcoal markets to make it economic to try to sell beyond the district. For example, in 1819 a wood near Killin was expected to yield 14,500 stones of bark, which at two shillings per stone amounted to £1,450. After deduction for cost of cutting, peeling and winning the bark, the net worth was estimated to be £1,150. In comparison, the return from the peeled timber was considerably less, at £150. Not surprisingly, Lord Breadalbane tried to reduce costs further by suggesting 'the tenants who want timber to take it for the barking'.

One of the inevitable results of coppice management for bark or charcoal, whether or not the wood was enclosed, was a reduction of species diversity among the trees. Oak was regarded as so superior in respect of the tannin quality of its bark, and the nature of the charcoal that it produced, that other trees were weeded out, and gaps filled up with new oaks grown from acorns, not all of them of local provenance. The result was to create an artificial monoculture of oak, fine and interesting in its own way, but not the same as the more diverse broadleaf wood that it replaced.

Caledonian Pinewoods

There is something awe-inspiring about our ancient Caledonian pinewoods. Even though they only cover a fraction of Scotland's remaining woodland area, this bewitching quality alone makes them deserve special mention.

Recently, the Forestry Commission discovered an old pine, which is at least 525 years old. It is one of several thought to be around 500 years old, among a scattered remnant, barely surviving as woodland. That tree, standing sentinel high up in its lonely glen, has endured through its long life a period of profound change, not only for woods, but also for people. By 1600, it was already around 150 years old, at a time when Pont and other early seventeenth-century travellers were beginning to describe the bountiful Highland woods. How did it manage to survive the subsequent 250 years of escalating exploitation of Highland pinewoods, as well as the last 150 years of neglect, further exploitation, and replacement by exotic conifers?

If it had been growing in the Glen Moriston pinewoods in 1624, when John Grant of Glenmoriston sold 'the number of auchteinscoir fir jeastis, all being fyne guid and suffcent reid wood' to the provost and baillies of Inverness, it surely would have been big enough to produce joists (beams). Of the 360 joists that were to be cut, and floated down to Inverness, half were to be 24 feet long and 15 inches wide, the other half, 24 feet long and 11 inches wide. A tree 150 years old might surely meet these requirements. But this remnant was higher up beyond Glen Moriston, within the royal hunting forest of Clunie. It was therefore too far from the river and loch to be worth felling. And 360 is not a vast number of trees in such 'great long woods of firr trees'. This relatively small-scale local trading in pine timber seems to have been widespread in Highland pinewoods from the early 1600s.

Always, in the intricacies of making money from Scottish woods, proximity to market was crucial. Coppice wood owners overcame this problem by concentrating on bark for more distant markets and timber for local consumption, unless the consumer came to the raw material, as with the iron-makers. Pine timber, as we see with the Glen Moriston contract, was generally tall and bulky; it didn't make good charcoal (too resinous) and it didn't coppice (put out new shoots after cutting). It was therefore less versatile than deciduous trees, and its value lay as a construction timber, producing beams, deals (planks) and spars (poles). Lochs and rivers were crucial for transporting this timber to the burghs that fringed the Highlands, or further afield. Overland transport of pine was not unheard of, but it would reduce profit.

The bottom line was that pine timber started with an inherent disadvantage over coppice wood. Not only could the Baltic furnish urban demand for construction timber, but it was not unknown even for Highland landowners to 'shop abroad'. When Inveraray Castle was being built in the eighteenth century, word was put around Norwegian captains that it might be worth calling into Loch Fyne on their way to the Clyde, where profit might be made from selling their timber cargoes to the Duke of Argyll. And yet, Inveraray is within easy sailing distance of the Ardgour pinewoods. Timber size may have been an issue and volume certainly was, because, just like today, Scandinavia had vast areas of forest in comparison to Scotland, producing greater lengths, widths and volume than Highland pinewoods.

Quality was also important, and may explain why the Duke of Argyll went to Norway, even though he was relatively close to a pinewood. The Earl of Aberdeen described the Abernethy pine as 'too knotty' when building Haddo House. The Duke of Gordon's architect described it as 'the roughest and coarsest kind', and a Fort William man turned his nose up at Loch Arkaig pine for a 'little washing house' because in his opinion foreign pine 'will be better timber and better season'd'.

Nevertheless, where Highland pine could be transported easily and cheaply, it was sold. The Inverness burghers could have as readily shipped in Norwegian pine or spruce in 1624, so they must either have squeezed a good deal out of Grant, or did not need high-quality timber – perhaps both. Narrow profit margins did not, however, stop the York Buildings Company trying to profit from Highland pinewoods. In 1728 they bought 60,000 pine trees from Abernethy. Some of these trees might have been intended for charcoal for their Abernethy ironworks (although brushwood would have been preferred) but most were to be floated down the Spey to Garmouth and then on to Newcastle.

The great innovation that this company brought to Strathspey was later remarked on by William Lorimer, who said, 'I don't find the York Buildings Company introduced any Improvements in manufacturing Timber, but that of Floating.' In fact, the change from floating small loads of deals and logs behind currachs, to much larger fabricated rafts, held together by spars and iron fastenings with men on top guiding them, was revolutionary. Designed rather surprisingly by a poet, Aaron Hill, who was

PLATE 5.4 *Glenmore south of the sawmill, 1786, from an original drawing by Charles Cordiner, minister of Banff. Many small mills processed Scots pine in the native woods. Building dams to create a rush of water that would carry logs downstream on their release could also lead to serious erosion, clearly seen here on the right. See also Plate 5.8 below. (St Andrews University Library.)*

working for the company, this innovation meant it became much cheaper to transport the timber to the coast. He also showed further imaginative flair by overcoming rocks impeding the floating operation by building big fires on them at low water and then throwing water on them to make them crack.

With this level of resourcefulness, it is a wonder that the company did not lead the field in profiting from Highland woods. Edward Burt, a soldier posted to the Highlands around this time, rather prophetically wrote of these Abernethy timber operations that no trees 'will pay for felling, and removing over Rocks, Bogs, Precipices and Conveyance by rocky Rivers, except such as are near the Sea-coast, and hardly those; as I believe the York-Buildings Company will find in the Conclusion'. How true those words were to become, for the company never completed its fifteen-year contract to fell the 60,000 pine from Abernethy and met with financial ruin around 1737. Burt was an engineer, involved in opening up the Highlands by improving the roads, so, ironically, it was the very task that he was engaged in that helped improve the viability of this trade, later in the century.

In Deeside, a different form of local trade existed. Here, it was not only the owners who were dealing in pine timber, but also their tenants, and very successfully too. In

PLATE 5.5 *The bucket mill at Finzean, Aberdeenshire, a nineteenth-century mill restored by the Birse Community Trust and still operating today in a traditional manner (above) from the outside (opposite) checking the finished buckets, 1958 (below) repairing the mill-lade, c. 1900. (© The Trustees of the National Museums of Scotland.)*

1725, a traveller described the local folk as 'living more by trafficking in timber than husbandry, this wood they have from the Wood of Glentanar'. Such trade continued into the 1760s, at least on the Mar estate, where it was noted that 'many of Lord Fifes tenants here spend a great deal of Summer in buying at the Woods large Boards manufactured and carrying them to Angus etc. and there selling them'. By then it was

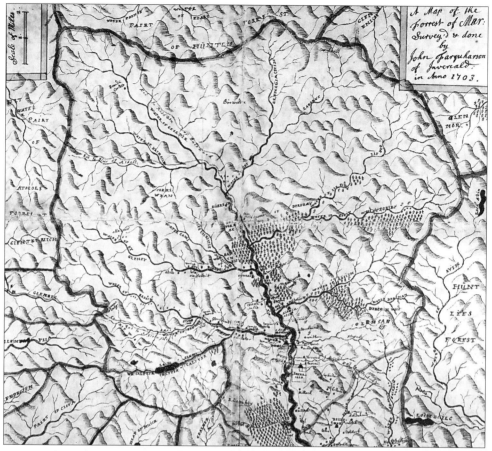

PLATE 5.6 *The Forest of Mar portrayed on one of the earliest Scottish estate maps in 1703. Most of the land was unwooded even then, but there was more wood, especially in Glen Derry, than now. (Reproduced by permission of the Trustees of the National Library of Scotland.)*

being frowned upon, as William Lorimer, who was advising a Grant laird in Speyside, went on to say, 'They spend too much time in this and neglect Agriculture.' Owners, it seems, were opening their eyes to greater exploitation opportunities, and might be better advised to cut out these traditional middlemen.

Despite occasional dealings with English companies during the first half of the eighteenth century, much of the trade in Strathspey followed similar lines to Deeside, with local men buying standing trees or sawn deals and so on, to be sold on, either locally, or to Inverness and the Moray coast. There was tremendous variation between pinewood areas in the form this trade took, but essentially it was local. However, the French wars did not leave the pine trade unaffected and as with the coppice woods, the period from the 1780s to around 1820 saw the greatest level of exploitation and change. Until then, pinewood owners had tried to interest outsiders in their timber and although the woods had furnished an amount of ships' masts and planks to the English navy and wooden water pipes to London, this trade never quite met the expectations of either seller or buyer. The war, however, brought disruption to European trade and an opportunity to profit for pinewood owners.

Most pinewood owners went headlong into increased production, with the number of sawmills and the volume sold both increasing. This period saw an intensification of activity rather than new forms of trade. Abernethy had four sawmills by the end of the century, and a pipe-boring mill. Some owners managed to sell their woods standing, like the Duke of Gordon who sold Glenmore forest for twenty-six years to an English company in 1783. 'Our ears are stunned by the crash of falling trees and clamours of the Sassenachs,' wrote a local bard of the Glenmore operations, but a frigate and several East Indiamen were built from the wood. The Black Wood of Rannoch was similarly sold to a public company in the early 1800s. However, most

PLATE 5.7 *A currach preserved in Elgin Museum. These circular boats of basketwork construction were used for attending the logs on the Spey, before the York Buildings Company introduced rafting around 1730. This was found at Mains of Advie in Strathspey and photographed in 1935 by B. Wilken. (Reproduced by courtesy of Elgin Museum.)*

to be followed by a period of protection. This came about in the mid-nineteenth century, at the end of a period of very heavy felling. There are few very old trees in Rothiemurchus, yet its continuation as an ancient woodland arises from the fact that regeneration was encouraged and allowed in the nineteenth century (see Plate 5.10).

By the time the Rothiemurchus woods were being fenced, the winds of change were already blowing through the Scottish countryside and towns. Scotland was rapidly becoming an industrialised society, with prosperity dependent on the developing manufacturing might of the central belt and the ability of its farmland to provide food for the ever-expanding urban masses. New processes and substances were developed, inventions created, fortunes made and greater accessibility to the colonies of the British Empire, old and new, brought cheap raw materials. These pressures for change would be irresistible and the land, its woods and its people would be irretrievably altered.

Changing Fortunes

When Victoria came to the throne in 1837, Scotland had changed beyond anything James VI, her predecessor of some 250 years earlier, would have recognised. He may not have approved of the deterioration in sovereign authority, but he would surely have applauded the prosperity of industrial barons and landed nobility. No doubt he would have smiled wryly at the final subjugation of the Highlanders, although tinged with bitterness, for this final blow to the Highland clans at Culloden was also the last hope of restoring the Stewart dynasty to the crown. The Scottish landscape was also much altered, and although James, in viewing nineteenth-century Scotland from his heavenly throne, might have bemoaned the throwing over of the old order, he would surely have been mightily impressed by the prospering and industrious towns, and farms that were the envy of Europe.

Our twenty-first-century landscape has its origins in the eighteenth and nineteenth centuries – the dykes, ditches and hedges of the Lowland countryside, the ruinous gable ends crowned by a solitary rowan tree in lonely Highland glens. These are the relics of the new order, created as Victoria came to the throne, when Scotland was a prosperous country, on the brink of great industrial advances. Conversely this was a period of considerable human suffering. Not just for Highlanders, whose story has become one of the most poignant of our history, but also in the Lowlands. Here, land became concentrated in fewer hands, leaving many landless, migrating to earn a living as hard-graft labourers, sometimes in great poverty. But how did our semi-natural woods fair in the early Victorian era?

We have seen how their value increased during the Napoleonic Wars, particularly the coppice-producing woods, but also, if less prodigiously, the pinewoods. If we value something, we look after it. But, people can be fickle, and if that value diminishes, then at best neglect will set in, at worst, it will be destroyed to make way for something of greater value.

Let us look first at Lowland woods. We have seen how they generally maintained their value throughout the seventeenth and eighteenth centuries, often carefully

PLATE 5.10 *Rothiemurchus forest at dawn. One of the greatest of Scots pinewoods in Strathspey, it survived very heavy felling in the first half of the nineteenth century. (Laurie Campbell.)*

managed, particularly close to towns, where profit was easily obtained. Occasionally they were lost to the needs of agriculture, usually where alternatives were readily available and profits not so lucrative. By the 1800s there was therefore undoubtedly less woodland than in 1600, but not significantly so. Woodland reduction and deterioration was probably most marked in more marginal areas like the Borders and Galloway, where sheep and cattle rearing took precedence over woodland management. Sometimes, though, as in the case of the early eighteenth-century Canonbie ironworks, a local demand encouraged careful attention to the woods, for a time at least.

Most markets for coppice produce were maintained and developed in the first half of the nineteenth century. After the war, home-produced tanbark kept its value for a while, but demand had always outstripped supply and consumers went elsewhere for their bark. The main challenge to the domestic bark trade came from imported oak bark, which could be more readily sourced after the war, so although tanning continued to be an important user of Scottish bark throughout the nineteenth century, the value it realised in the early 1800s would never be repeated.

As for coppice timber, while some industries switched to coal and coke, notably the iron industry, others developed that consumed huge amounts of coppice wood. The textile industry (encompassing spinning, weaving, bleaching, printing and dyeing), one of the great Scottish industrial successes of the nineteenth century, provided fresh demand for coppice timber, in various forms. From wicker baskets for carrying cloth, to mill machinery and bobbins, which were manufactured throughout Scotland for many large- and small-scale operations, and, perhaps most demanding of all, for the production of acetic acid. The latter took the form of the distillation of wood, leaving tar, charcoal and vinegar as residue, the resulting acid being used to produce mordants for dyeing. Near Perth, for example, by 1845, a large textile company was operating a 'pyroliginous' plant from which charcoal, as a by-product, was sold to foundries and dyeworks in the city. The Vale of Leven was also a hot spot for textiles and for gun-powder production, which used alder in particular, but also birch, hazel, willow and rowan. The semi-natural woods of the parish of Buchanan, including those of the east side of Loch Lomond and new oak plantations to the south, supported a pyroliginous works at Balmaha until the 1920s, producing acetic acid and dye-liquids for Glasgow printing works.

Other more traditional industries continued to use coppice timber, such as cooperage. Although imported timber was more popular, there was still a market for home-produced barrel staves, which when cut at source were more portable and therefore could be an attractive commercial proposition even in more remote areas. Industries relying on barrels included the more obvious drink manufacturers and also the herring and salmon industries. Domestic use of barrels, kegs and buckets remained important local users of coppice produce, not only for making staves, but also the hoops, although hazel and willow hoops declined in the industrial sector, replaced by iron.

As the Scottish road system improved, more and more carts, carriages and other wheeled vehicles made their way on to the roads. This required wood for spokes and small coppice timber fitted the bill exactly. This used timber as small as 3 inches in diameter and up to 2 feet long – if worked at source then this could be much more easily carried to the nearest coachworks. By the 1860s, bark and spokeswood were the most valuable part of a coppice.

Together these uses of coppice produce offered sufficient encouragement, in some areas at least, to maintain dykes and fences round woods. It also prompted woodland owners to continue to alter the composition of their woods, so that they could increase their production of the more valuable oak. This went hand in hand with a more professional approach to estate management and the development of plantation forestry, guided by an increasing band of silviculturalists and forestry advisers. Not only were 'vacancies' filled with oak, but in some woods, the result of whim as much as of economic expediency, exotic trees were also planted, such as beech (native only to the south of England), Norway spruce and larch.

The great trend of the late eighteenth and early nineteenth centuries, linked to the drive for agricultural 'Improvement', was tree planting. This took various forms in different parts of the countryside and will be examined in more detail in the chapter that follows. Nevertheless, this drive for planting in semi-natural woods, whether to provide ornament, cover for game or additional profit, has left a legacy for twenty-first-century woodland managers. Many of these former coppice woods retain a greater proportion of oak than nature intended, like the coppice woods of Buchanan, or the Strathtay oakwoods, where they survive, almost completely dominated by uniform stands of oak.

Other woods became incorporated into the newly enhanced, designed landscapes of mansion houses. Often they centred on pretty waterfalls, where hermitages could be built, at the end of a fine carriage ride from the big house, and were 'adorned' with interesting newly introduced trees. Nature simply was not picturesque enough, or as William Marshall, one of the 'Improvement' advisers, so revealingly put it, 'Our idea of "natural" is not confined to "neglected" nature, but extends to cultivated nature, to nature touched by art, and rendered intelligible to human perception.' This woodland 'cultivation' has left many of our most important semi-natural woods like those of the Clyde Valley, the Lothian and Fife dens or the Ayrshire and Border cleughs with formidable management problems, linked to the invasiveness of some of these trees and shrubs, not least beech, sycamore and rhododendron.

As for the Highland woods, some of the same influences and pressures were at work as in the Lowlands. Not surprisingly, those coppice woods that were close enough to tanneries, coachworks and pyroliginous works continued to yield a good profit and were maintained with supplemental planting, just as in the Lowlands. Some, like the Killiecrankie woods, which provided the additional non-commercial benefit of scenic tourist views, also succumbed to 'adornment' with beech, larch, spruce and fir. The Highlands were becoming an increasingly popular place for travellers, keen to view the romantic untamed landscapes, as Sir Walter Scott would have it, 'so wondrous wild'. There could occasionally arise a dilemma here for landowners, who wanted to continue to exploit their woods, but did not want to risk a literary dressing-down for spoiling their estate's picturesque assets. One answer, as the Earl of Breadalbane suggested in 1814, when deciding on the felling of a group of oak trees, was to ensure that 'as the old oaks are mostly out of sight of the Public Road, the appearance of this beautiful Glen will not in the smallest be hurt in view of the many travellers who frequent in the summer season'.

For most Highland woods, however, their fate was much worse than enduring the planting of a few ornamental trees. We have seen how increasing dependence on black cattle had already put pressure on the woods towards the end of the eighteenth century. This was exacerbated by an unprecedented rise in the Highland population. With the end of the Napoleonic War, cattle prices fell, and it became more difficult to compete with more productive Lowland cattle breeders. The Highlanders clung on to their traditional way of life, and even in the early twentieth century, it was common for Highland townsfolk to keep a cow. But, their lairds and chiefs were no longer in need of large populations of fighting men, nor were they content with the old ways. They became landlords, divorced from their cultural roots, educated in England or abroad, and used to the finer things in life. Some were very amenable to change, others were cajoled by their advisers into change, particularly if it realised greater personal wealth. And change came.

The coming of the sheep and the Highland clearances remain emotive subjects for Highlanders, both at home and abroad. If the redcoats' musket had shattered Highland culture, then the Lowland sheep tore their traditional economy out by the roots. The woods, some already showing signs of exhaustion from too many cattle, had little hope of bearing a further onslaught from thousands of hungry mouths. Pressure of numbers was to be simply too intense to allow woodland regeneration.

The expansion of large-scale sheep farming was gradual in the Highlands, being first introduced to the southern Highlands in the 1760s, gradually making their way north and west over the following fifty years. To begin with only the upper glens and more marginal parts of estates were taken over by sheep, but these were often the same glens where the hill woods had already become exhausted, like those of Glenorchy and the Forest of Mamlorne. Only the ability of pine to grow to a great age, or the rowan to cling on to rocky crags, or the birch to withstand nibbling mouths, allowed scattered remnants to remain through the nineteenth and twentieth centuries. For many, it was not death by a thousand cuts, but death by a thousand mouths.

Woods lower down the glens, along lochshore and riverbank retained some value and only gradually, as the sheep steadily grew in economic importance, did the dykes start to crumble and not to be repaired. It is an interesting twist of fate, that the most notorious landlord's henchman of the 'clearances', Patrick Sellar, the Sutherland estate factor, in 1816, just two years after his infamous evictions of Strathnaver, demonstrated his esteem for woodlands over people. He made a survey of the estate's woods, looking for ways to profit from them. He noted in Strathnaver that the woods had 'already begun to shoot out since the removal of the tenantry. That part thinned last Year, by the [incoming] tenant, shews a new appearance of health.' The health of the old tenants was not a great concern, and if any of them were caught damaging the trees, they too would be evicted.

These woods may for a time have responded to the new order, most probably because there were fewer people using and abusing the woods than previously. The Sutherland estate alone expelled between 5,000 and 10,000 people between 1807 and 1821. The brief respite was not to last. Furthermore, it was not only the remote, thinly wooded parts of the Highlands that became sheep runs. The pinewoods of Glentanar in Deeside were said to have 7,000 sheep running through them by the middle of the nineteenth century.

As some woods were opened up again to stock, others continued to be closed, not only to animals, but to humans too. This protection was not for the trees, but for the game that they sheltered. The chiefs and lairds, in their new guise as landlords, had regained their taste for hunting and sport. Victoria and her hunting-mad husband added momentum to the revival in field sports, and the newly prosperous industrialists of the south, the ironworks and mill owners, headed for the Highlands, buying up the estates of debt-ridden clan chiefs and employing the remaining population as ghillies and stalkers. To an extent, this was a double-edged sword, although one side was sharper than the other. Lowland woods and those in the Highlands valleys were protected for game. The dykes were kept up and people were kept out. The collecting of bracken, which had long been cut from woods for thatching, was forbidden in such woods, in case the dykes were damaged and the game disturbed. At the same time, however, game cover such as rhododendron was planted and is now the scourge of western woods.

Sheep were never going to be the panacea Highland landlords and their advisers hoped for, and after the middle of the nineteenth century, many of them happily became sporting landlords. Once again, the hills were turned over to deer forests, except that by the 1850s, few of these forests had many trees left. Most of the pinewoods became deer forests, and as the deer populations increased, there was little chance of much regeneration. Grouse too gained in popularity, and muirburn, once a practice completely frowned upon by landowners for the damage it would do to trees, particularly pine, now dominated the management of moors, again to the detriment of any seedlings that showed their heads above the heather, in the periods between burning.

If, occasionally between 1600 and 1850, the value of trees and woods outweighed that of animals, when demand and return for woodland produce – bark and charcoal – were higher than for animal products, nevertheless, animals invariably won through, in an essentially pastoralist society. Therefore, while occasional battles were won over the years, the war was inevitably lost. It was a war that had been raging since people first came to Scotland with their animals, but what was perhaps unique about this period, was that not only did the semi-natural woods lose their value, but so did the people. We were left, at least in the Highlands, with only a few lairds and their factors in direct command of the land, its people and its woods.

CHAPTER SIX

'A Nation of Planters':
Introducing the New Trees, 1650–1900

SYD HOUSE and CHRISTOPHER DINGWALL

There is no epithet, by which the inhabitants of the Northern Division of this Island in the present day, can be more appropriately distinguished, than that of a 'Planting Nation', or to speak with more correctness, a 'Nation of Planters'.

Sir Henry Steuart, *The Planter's Guide* (1828)

INTRODUCTION

For a country once predominantly covered by natural woods, Scotland had become a country largely barren of its tree cover by the start of the seventeenth century. Large tracts of surviving woodland remnants still existed, particularly in the remoter Highland glens, but the lack of woods and trees often drew comments from travellers and visitors, including the renowned Dr Samuel Johnson, who wrote in 1773 that 'a tree in Scotland is as rare as a horse in Venice'. Johnson was, of course, not entirely objective in his view.

The reasons for this lack of tree cover have already been explored. A combination of clearance for agriculture, overexploitation of the timber and consequent overgrazing by domestic stock, allied to ineffectual forest protection, were primarily to blame. The 'natural' openness of much of Scotland's contemporary landscape is in fact a result of the enormous impact human design has had. Virtually all the natural woodland has disappeared over the centuries, to the extent that today only around 1 to 2 per cent of Scotland is still covered by the much altered remnants of the natural woodland cover – a level of forest clearance more effective and widespread than in any tropical country.

Despite this neglect, there had been since the time of the monasteries in the Middle Ages, an active interest in the management and planting of trees. The introduction, for example, of many of the fruit trees and cultivars we think of today as traditional to Scotland was one such outcome of the monks' efforts. From these early, localised beginnings, and the inauspicious circumstances facing Scotland's woodlands in the seventeenth century, was to grow, in the eighteenth and nineteenth centuries, a burgeoning interest in tree planting and forestry. Ultimately this was to lead to the introduction of new species of trees from throughout the world, as well as the development of plantation techniques for establishing new woodlands on bare land even to the extent of completely replacing native woods by new introductions.

PLATE 6.1 *An old sycamore at the Loch of Clunie, Perthshire, growing on a former castle mound. It has clearly been pollarded at one stage, and its great size helps to confirm that sycamore has been grown in Scotland from at least the sixteenth century. (Anne-Marie Smout.)*

PLATE 6.2 *Large old sycamores are not new: this was drawn by J. G. Strutt at Bishopton, Renfrewshire, in 1822, and measured 20 feet in circumference and 60 feet in height. David Alderman of the Tree Register of the British Isles considers that such a tree would have originated in the fifteenth century. (St Andrews University Library.)*

The story of Scotland's modern forests and woods, the trees that grow in the forests and the pattern of forestry management that define the way they fit in the landscape today, has its origins in the first major efforts at large-scale tree planting in the eighteenth and early nineteenth centuries. This chapter is about that 'nation of planters', the drive for the establishment of new woods and the renaissance of forest management.

THE FIRST ATTEMPTS AT FOREST MANAGEMENT, 1600–1725

In the seventeenth century, in recognition of the serious lack of timber and woods, allied to an increasing interest in 'improving' land management and landscape enhancement, a number of new attempts were made to promote tree planting and the establishment of new woods. In 1616 an Act of the Scots Parliament encouraged 'civil and comelie' houses with 'policie and planting about them', while in 1661 Charles II passed an Act whereby:

Every Heritor, Life-renter, and Wodsetter...within this ancient Kingdom of Scotland, worth one thousand pounds of yearly valued Rent shall inclose four Aikers of Land Yearly at least, and enclose the same about with Trees of Oak, Elm, Ash, Plain [sycamore], Sauch [willow], or other Timber.

In fact the law simply confirmed a trend already under way. The landed classes were already busy surrounding their homes with trees as suggested by the maps of Timothy Pont, compiled just before 1600, which show a greater extent of wooded parkland and tree clumps around great houses than previously thought. In the course of the seventeenth century the large land holdings of the crown were broken down and sold on to smaller landowners keen to effect improvement to their properties, while new ideas such as those outlined in John Evelyn's hugely influential book *Sylva; or, A Discourse of Forest Trees and the Propagation of Timber*, published in 1678, made a significant impression on potential 'improvers', including his great friend, the Earl of Tweeddale, who was an enthusiastic planter at Yester in East Lothian.

The rationale behind these early attempts was partly financial, in terms of getting a timber crop, partly agricultural improvement in the form of shelter for crops and

PLATE 6.3 *An avenue of beech shading the main road from Perth to Inverness, north of Dunkeld, in 1876. Beech was introduced for its combination of beauty and utility, and was often 'bundle-planted' (several stems together) to achieve an impressive buttressed trunk quickly. The tree on the right was already 15 feet in girth. (Reproduced courtesy of the Royal Scottish Forestry Society.)*

animals, and partly amenity. These early attempts at landscape planting were heavily influenced by European styles, particularly French gardens like Versailles typified by rectangular patterns with straight avenues and criss-crossing rides cut through the woods. Drummond Castle and Taymouth in Perthshire, Glamis Castle in Angus, Haddo House and Pitmedden in Aberdeenshire, Drumlanrig in Dumfriesshire, and Kenmure Castle in Kirkcudbrightshire were all examples of Scottish emulation of their Continental peers.

Sir William Bruce (1630–1710) was one of the most important architects of the time, whose use of trees was to influence many of his fellow gentry and landowners in their attempts at landscape design around their own great houses. Often called 'the Christopher Wren of the North', he remodelled Holyrood and Hopetoun House as well as his own home, Kinross House, putting great emphasis on the relationship between the house and the landscape, including the use of trees.

Native species such as oak, ash, elm and Scots pine were the principal species used in these early experiments, though often, with Scots pine in particular, these were established on ground outside their former natural range. There was also a willingness to experiment with introduced trees from Continental Europe or England such as beech, sweet chestnut, sycamore (commonly called 'Scotch plane' at this time), Norway spruce, larch and European silver fir, none of which were native in Scotland, yet which produced valuable timber that the 'improvers' felt could be grown at home. Some of these trees, notably Norway spruce, beech and sycamore, had probably been introduced into Scotland much earlier and were to thrive in northern latitudes.

The spread of learning and advice was helped by the founding in the 1670s of the Physic Garden in Edinburgh, later to become the Royal Botanic Garden. A wide range of introduced trees was listed in Sir James Sutherland's first catalogue of the plants growing there in 1683, including the European larch, well before the more famous Atholl larch introductions of the next century, as well as other European trees, together with trees native to eastern North America, which was then in the process of being opened up for settlement.

In 1683 the first publication dealing with woodland trees was published in Scotland. *The Scots Gard'ner* was written by John Reid, who was employed by Sir George Mackenzie of Rosehaugh, though he worked on various estates across Scotland. The book was to become a landmark influence on future gardeners, tree planters and 'improvers'. In his book Reid recommended how and where to get tree seed, how to propagate trees from seed, the choice of trees available, both native and introduced, the treatment of pests and diseases, and other silvicultural advice. Reid exhorted tree planting 'for profit and pleasure', neatly encapsulating much of the spirit behind the tree-planting endeavours of the age.

By the end of the seventeenth century these experiments were to prove vital, given the long-term nature of forestry, in providing a period of trial and observation essential to the great 'Planters' of the eighteenth century. This was a pattern which would repeat itself – the introduction of a new tree as an ornamental, subsequent trial and observation in larger plantings, leading, if appropriate, to full-scale forest plantations, even to the extent of overusing the tree and planting on unsuitable sites.

PLATE 6.4 *Blair Adam, Kinross-shire, in 1834. Policy planting was carefully designed to help focus the eye on distant features, here on Loch Leven castle and Benarty hill. From an anonymous artist and engraver. (Christopher Dingwall.)*

If successful, there would be a clear place for the tree in the silvicultural options open to the forester and landscape gardener, or alternatively it would revert to one of the considerable numbers of minor species capable of being grown in Scotland and of interest to the landscaper, botanist and arboriculturist. One century's trial would often be the next century's favoured tree. Foresters in Scotland, whilst having a limited choice of native species to work with, have been compensated to a large degree by the great range of introduced trees that grow well in Scotland and that outperform native species in terms of growth and productivity in sites of limited fertility.

One of the most interesting insights into the contemporary approach to tree planting comes from the rivalry of the Earls of Argyll and Atholl. The 9th Earl of Argyll was an enthusiastic tree planter, getting advice directly from John Evelyn on what trees to plant around Inveraray Castle, a fact which Evelyn acknowledged in later editions of *Sylva*. Such enthusiasm was to have painful consequences for the Earl, caught up as he was in the various power struggles and political machinations of the time. In 1685 the Murrays of Atholl, temporarily responsible for the estate and the grounds of Inveraray Castle during one of the many periods of political instability, could not resist the temptation to plunder the grounds and removed some 34,400 live trees including sizeable saplings of silver firs, pines, beech, lime, walnut and fruit trees to Perthshire. The logistics required to transport such booty to Atholl, in addition to the well-known sensitivities of live plants when handled less than carefully, all raise questions as to the real purpose of the deed, yet the story is well recorded in Atholl history. There do not seem to be any other instances of large-scale tree rustling, so it

PLATE 6.5 *Allanton House, Lanarkshire. In planting his park in the fashionable picturesque style in 1816 and 1821, Sir Henry Steuart often used mature trees, which he moved with his 'transplanting machine' on the left of the picture. From an engraving by W. Miller after a drawing by W. Turner. (Christopher Dingwall.)*

can only be assumed that the cost of rearing and growing these trees made them a significant prize for the tree enthusiasts of Atholl.

What effect did these pioneering efforts have? Locally, at least, they were significant. In contrast to the many travellers who noted the lack of trees in Scotland the celebrated writer and diarist, Daniel Defoe, writing his *Tour* in the early part of the eighteenth century, observed:

> You hardly see a Gentleman's House, as you pass the Louthians, towards Edinburgh, but they are distinguished by Groves and Walks of Firr-trees about them . . . In a few years, Scotland will not need to send to Norway for Timber or Deal, but will be sufficient of her own and perhaps be able to furnish England too with considerable Quantities . . . Improvements are already of 50 to 70 and 80 years standing as at Melvil, Leslly, Yester, Pinkey, Newbattle, and several other places.

Defoe was perhaps a little early in his vision of Scotland becoming a net exporter of timber, though that situation is upon us some 300 years later.

Even more importantly, this early interest was to pave the way for the first great wave of planting that was to flourish from the middle of the eighteenth century.

Social conditions were changing. Following the establishment of the union with England in 1707, trade and wealth gradually increased. New crops and animals were introduced in the general drive for 'improvement', creating a state of mind among estate managers which sought to make every piece of land useful for some purpose or other. This is hardly surprising, for Scotland had suffered terribly in the latter part of the seventeenth century from famine. Hard times are great motivators for change and improvement. For many landowners, the existence of so much bare and unproductive 'unimproved' land offended their work-and-profit ethic and the God-given right to make all land serve the purpose of humankind. Robert Burns in 1788 called the stretch of hill land between Galloway and Ayrshire, 'A parcel of damn'd melancholy, joyless muirs', a comment typical of the age. The thought of using such land to grow productive timber would have seemed as natural as planting the recently introduced potato, a revolutionary agricultural introduction which would subsequently enable the population in the Highlands to expand rapidly.

The Planting Movement, 1725–1830

The eighteenth century and early part of the nineteenth century was a momentous time for Scotland, with major impacts on the country's cultural life, and economy, on the creation of wealth and the productivity of the land. This was 'the Age of Improvement' and the 'Scottish Enlightenment'.

From the perspective of tree planting this was also to be a 'Golden Age'. The great 'improvers' were radically changing the face of the land. Allied to the establishment of learned bodies and societies to share this new knowledge (a legacy borne out of the Scots' investment in education), these forces together were to encourage a drive to arrest the long history of forest decline and to initiate large-scale new woodland establishment. Inevitably there was increasing interest in using introduced trees such as larch to improve the economic return from tree planting. Such a willingness to innovate was to reach a peak in the nineteenth century, a recognition of both the drive of landowners for increase in productivity, and that the choice of native species was very limited compared to elsewhere.

But not everything was for utility. 'In my opinion Planting ought to be carried on for Beauty, Effect and Profit' wrote 'Planter John', the 4th Duke of Atholl (1755–1830), writing in his forestry journal and putting profit third (see Plate 6.6). This was also an age of the great landscape designers such as 'Capability' Brown and Humphrey Repton, whose influence was to stretch into Scotland. Gardens and the landscape around country houses were now treated as an extension and integral part of the neo-classical architecture which was becoming increasingly popular as many of the gentry undertook the 'Grand Tour' of the great cities and classical sites of Europe. These early tourists also viewed and collected the pastoral images and paintings of Italian landscapes, creating a desire to emulate what they had seen, in an English or Scottish setting. Roy's comprehensive maps of mid-eighteenth-century Scotland show the early beginnings of the great range of planting and trees around the houses of the landed gentry such as at Taymouth Castle, Blair Adam and Inveraray Castle.

PLATE 6.6 'Planter John', 4th Duke of Atholl, 1755–1830. He hoped to supply the entire British navy with larch from his estate. (Courtesy of the Atholl Estates, Blair Castle.)

With even light fortification no longer important and greater wealth flowing into the country, landowners were keen to build new houses and beautify their surroundings. The earlier formal style, such as at Glamis Castle or Haddo House, evolved to become much more informal, with the 'parkland' style so characteristic of places such as Taymouth. One of the most influential of the school of landscape designers working at this time was Thomas White (one-time associate of 'Capability' Brown), and his son, also Thomas, who established a successful practice in Scotland. White prepared improvement plans for some seventy estates throughout Scotland, mainly in the Lowlands, though Gordon Castle and Cullen House in the north-east were exceptions. He had a distinctive planting style and was 'very partial to larch' as he said of himself. His work and aims at Scone, described in a letter to Lord Stormont written in 1784, outlined his philosophy:

> To raise up such plantations as will shut out its deformities and heighten its beauties, and by breaking the distant scenery as you pass along, introduce the same objects in different lights and points of view and thereby give you an agreeable change of scene.

Sir Henry Steuart, author of *The Planter's Guide* in 1828, described Thomas White as 'an ingenious planter' who was able to combine aesthetics with timber production by using a high proportion of Scots pine, Norway spruce and larch to maximise the economic return from these new woodlands.

The story of the famous Adams of Blair Adam, an estate which straddles the border of Fife and Kinross, highlighted the philosophy of planting and landscaping over the period. When the estate was acquired in 1733 there was reputed to be only one single tree on it. By 1792 some 1,144 acres of woods, around a third of the estate, had been planted. William Adam, younger, in *Remarks on the Blair Adam Estate* (1834) talks of the different values of trees for economic and amenity objectives. On the ground this was translated into management of 'woods of succession', which were managed principally as a timber crop to be harvested at the appropriate time; 'woods of selection', where the object was to make a return yet maintain a continuous cover to retain a woodland environment in the landscape; and 'woods of policy and ornament', which were managed primarily to enhance their visual and recreational value. Anyone familiar with current forest design philosophy or who has been involved with the preparation of a long-term forest design plan will be entirely familiar with the sort of zoning approach described for Blair Adam nearly 200 years ago.

It was a feature of these early tree planting initiatives that landowners and their advisers were readily able to combine planting for financial gain, landscape enhancement and agricultural improvement. They saw nothing contradictory in accommodating each of these interests to their mutual benefit. However, the view from those lower down the social order, the peasantry and farming tenants, was nothing like as supportive. The nature of land holding in Scotland, with its roots in medieval feudalism whereby trees and woods were generally retained by the landowner, meant that the mass of people viewed trees as at best a waste of land and at worst part of some conspiracy to make life harder for them. In *The Life of Robert Burns*, Catherine Carsewell writes of Kincardine and the Mearns in the early eighteenth century in the time of Burns's father and grandfather, who were tenant farmers in the area:

In many places the bare and boggy countryside began to show some cultivation other than the savage kind so long unchallenged. Hedgerows – a happy inspiration from Holland – were set protectively here and there. Plantations of tiny larch seedlings, foot-high spruce firs, sapling beeches, oaks and elms, were being enthusiastically laid out and tremblingly watched over by the awakened landlords. If not grubbed up in the night by poor neighbours accustomed to treelessness and convinced that the roots would rob the soil of all nourishment, they would one day provide sheltering woods.

Sir Archibald Grant of Monymusk in Aberdeenshire was one of the earlier great tree planters who met resistance from tenants who cut down young trees and pastured stock to graze down new woods. Similarly James Hogg, the Ettrick Shepherd, would later describe the Southern Uplands as 'hills [which] formed such excellent ranges for sheep pasture that plantations of any great extent would have seemed in the eyes of the farmer a grievous encroachment'. Despite such resistance, landowners were to undertake throughout Scotland major projects to establish new woodlands. Their efforts were greatly helped by the establishment of many new societies. The Honourable Society of Improvers in the Knowledge of Agriculture in Scotland, founded in Edinburgh in 1723 and including such luminaries as the Duke of Atholl and Earl of Breadalbane, promoted new ways of managing the land, including tree planting. The Society for the Importation of Forest Seeds formed in Edinburgh in 1765 imported three collections from eastern North America in its first year. In 1786 the Highland and Agricultural Society (later to become the Royal Highland and Agricultural Society of Scotland) was founded with regular articles on forestry contained in its *Transactions* and out of which was subsequently to grow the Royal Scottish Forestry Society (RSFS). Such enthusiasm to spread good practice was readily acted upon. The pen of Sir Walter Scott, friend of the 4th Duke of Atholl and knowledgeable forestry enthusiast in his own right, was wielded in the cause of the new forestry and the RSFS was to use a quote from one of his novels as its motto: 'Ye may be aye sticking in a tree; it will be growing when ye're sleeping.'

Interestingly there is one early example of a 'community' forest being established in contrast to the more unilateral approach of landowners. Many Scots burghs held common land in the form of a burgh muir where local towns people could graze livestock and grow crops. In 1714 the burgh council of Perth agreed 'unannimously . . . to enclose and plant the said burrow muir' on its 340-acre holding for much the same reasons as the lairds, 'the planting of trees may prove both profitable and pleasant', though in this instance it was the 'common good' rather than a private enterprise that was the intended beneficiary. The muir was subsequently planted up (mainly with Scots pine) and over the next ninety years or so managed as a productive woodland with variable profitability, though the public amenity afforded the people of Perth was much appreciated. The woods were eventually felled in the early part of the nineteenth century, bringing a profit to the council of £13,352 10s 2d. The land was then converted back to farmland on the same basis as it had been originally planted – the forecast financial return.

THE GREAT PLANTERS OF THE EIGHTEENTH CENTURY

Perthshire holds a special place in the development of Scottish and British forestry. To many professional foresters it is known as the 'cradle of the Scottish forest renaissance' chiefly on account of the work of the 'Planting' Dukes of Atholl. Over the hundred years from the early part of the eighteenth century until 1830 the 2nd, 3rd and 4th Dukes were to plant over 21 million trees on some 15,000 acres of ground. Almost two-thirds of this area were covered by the European larch, a tree which has become strongly associated with the Dukes. Judging by their earlier willingness to relieve the Earl of Argyll of his arboricultural treasures, there had always been a healthy interest in new trees by the Dukes of Atholl. It was the larch, however, which was to make their name famous in the early development of forestry in Scotland.

Although larch had been grown as a specimen tree in Scotland for some time prior to its use by the Dukes, it was they who were bold enough to see its potential and utilise it on a large scale. Gifted some sixteen larch trees by a neighbouring landowner, the trees were planted out in 1738 next to Dunkeld Cathedral (see Plate 6.7) and Blair Castle, where surviving individual trees still grow today, and flourished to such an extent that they were to encourage major tree-planting efforts under the 4th Duke. In fact this and other large-scale efforts elsewhere in Scotland were the first significant

PLATE 6.7 *Of the five seedling larches planted near Dunkeld Cathedral in 1738, one was felled in 1789 to make mill axles, and two more in 1809, one being sent to the Thames dockyards for Admiralty trials. The other two became known as the 'parent larches', photographed here in the later nineteenth century. One perished in a lightning strike in 1905. The last one survives, over 100 feet high. (Courtesy of the Atholl Estates, Blair Castle.)*

attempts anywhere at establishing major plantations of conifer trees, as opposed to the management of modified natural forest in Continental Europe.

John, the 4th Duke of Atholl, was the most important and influential figure. He observed the properties of the larch, excellent for shipbuilding and general estate work such as fencing and housebuilding, and estimated that he could supply the Royal Navy with much of their timber requirements at significant profit, of course, to himself. The availability of cheap imported timber and the move to iron and steel for shipbuilding were later to put paid to his vision. Nevertheless the legacy of the tree-planting efforts of the 'Planting' Dukes is with us today, both in the superb forested landscapes between Dunkeld and Blair Atholl and in the inspiration to others.

Atholl was not alone. With the experience and observation of new tree introductions and silvicultural techniques in place, and with new publications available such as the 1761 edition of the 6th Earl of Haddington's *Treatise on the Manner of Raising Forest Trees* (one of the most influential books on tree planting of the period), other landowners throughout Scotland became significant tree planters. The most enthusiastic and notable of these were in Argyll, Perthshire and in particular in the north-east, where Archibald Grant of Monymusk planted some two million trees by 1754. In 1800 one commentator remarked that the laird of Invercauld had planted 'vast ... woods and firs ... in incredible numbers ... Few proprietors have done more ... toward the improvement of their estates than Mr Farquarson [who has] planted no less than 16 million firs and two million of larch.' The Earl of Fife at Duff House near Banff, the Earl of Moray on Darnaway estate and Seafield estate between them planted millions of trees in their drive to 'improve' their landholdings, one of the reasons why Moray and the adjacent counties became, and remains, one of the most heavily forested parts of Scotland.

In the case of Darnaway much of the planting was with Scots pine collected locally. Between 1767 and 1810 some 10,346,000 'Scotch' pine was planted out from seed collected on the estate, probably from remnants of the natural Caledonian pinewoods, illustrative of the long and important tradition of collecting Scots pine seed from the pinewoods of the north-east, either for use in the nursery for planting out or for export to other parts of the country. The Earl of Haddington remarked on this regular trade in the export of pine seed from the Highlands, while the Highland Society was later to offer premiums for collecting Scots pine seed. Even as early as 1621 James VI had asked the Earl of Mar to send seed from Mar forest for use in England.

The pattern of planting was repeated elsewhere. By 1750 there were over 700 acres planted at Drumlanrig Castle in Dumfriesshire, while in Perthshire, Breadalbane, Scone, Blair Drummond (where Lord Kames was to plant on the mosses reclaimed from the Carse of Forth) and Drummond Castle were establishing that county's distinguished tree-planting reputation. In the Highlands, the work of the Board of Commissioners of the Forfeited Estates, of which Lord Kames was a member, acted positively in encouraging woodland management and planting in places such as the Lovat estate.

In Argyll, the Dukes continued to be enthusiastic 'collectors of exotics' and planters, so that it was said that a horseman could ride for three miles up the River Aray through plantations. In 1756 the 3rd Duke, according to estate records, planted some 29,657 trees, which included a considerable number of introductions from the east coast of North America, such as 'American larch, 80 red cedars . . . 67 foreign oaks, 38 New England pines . . . 55 Carolina cherry . . . and 17 tulip trees'. The estate also promoted tree planting by other landowners, sending silver firs, larch and Norway spruce to Blair Adam in 1751 among other places.

In the Lowlands the pattern of planting was on a smaller scale and followed the pattern of agricultural improvement as exhorted by *The Scots Gard'ner*, which had laid out a profitable formula for oak coppice with standards which appeared financially very attractive, predicting a return of some £260 10s 6d after twenty-five years with the standard oak over-storey still standing. No wonder many landowners were enthusiastic tree planters. Many of the woods created in this period can still be found today in places such as Ladybank in Fife, which is surrounded by pinewoods originally established by the Melville estate on infertile sandy heathland, and Glamis in Angus, which had around 1,000 acres of plantations by the 1790s. Montreathmont forest, between Forfar and Brechin, was typical of this type of woodland development. Surrounded by fertile arable ground, Montreathmont was an unproductive wet Lowland heath, impossible to drain for arable crops or even permanent grass. The obvious thing for the estate to do to make it productive was to plant trees, and it has been continuously forested ever since.

Throughout this period, tree planting was remarkably widespread across Scotland with the possible exception of the Western Highlands. Although understandably only tried and tested species were generally used for large-scale planting, there was also a general drive to experiment with other species from wherever they could be found. Trees used for woodland planting were principally native species such as ash, alder, elm, oak, birch and, increasingly, Scots pine. Beech and sycamore were also extensively planted for timber, shelter and amenity, particularly in the eastern part of Scotland, where beech shelterbelts continue to be prominent in the landscape. Other broadleaves such as lime were used to a much lesser extent and generally with experiment and amenity in mind.

By the mid-eighteenth century smaller landowners were beginning to imitate their grander neighbours, a trend clearly seen on Roy's military survey. Encouragement was also given by an increasing number of books offering advice, among them William Boutcher's *A Treatise on Forest Trees* (1775). The list of some 400 subscribers to this publication includes an impressive cross-section of Scots nobility, gentry and landowners. Also of interest is *Miscellaneous Observations on Planting and Training Timber-Trees*, dedicated by its anonymous author to Lord Haddo (1777), which compares the productivity of broadleaved and coniferous plantations in some detail. The effects of this planting were subsequently to be described in the parish descriptions collected by Sir John Sinclair in the *Statistical Account of Scotland*, published in the 1790s.

In terms of the types of trees being planted, the most noticeable development was the increased use of Scots pine (often called 'fir' or 'firr') for plantations for timber as well as amenity, even on to sites where it was not entirely suited. After 1750 the use of

Scots pine expanded even more, as it was often planted on to bare heathland in mixture with other species. The planting of a wide variety of species was quite commonplace, as tree planters combined experiment with safety in terms of keeping their options open as to which trees were best suited to the ground being planted. Conifers were often planted as a 'nurse' crop with broadleaves, in order to ensure both that the broadleaves grew straight and to provide a cash crop at an early stage from the first thinning onwards, a practice still recommended (but underutilised) today.

Interestingly, as has already been described for Scots pine, there was a thriving trade in seed and plants, including imports from England and the Continent. Pedunculate oak (*Quercus robur*) was often preferred to sessile oak (*Quercus patrea*), which is naturally more common in the west and in upland areas. Estate records and newspaper advertisements provide evidence of a considerable trade in plants and seed. Commercial nurseries became very active during the period and were able to supply trees or seed from English, Continental and North American sources to meet the growing demand. The most famous of these was Dickson of Hassendeanburn in Teviotdale, which had been founded in 1729 and which was later calculated to have supplied enough trees to plant up to 48,000 acres.

Of the introduced conifers, larch was by far the most important introduction, becoming increasingly popular as a result of the Atholl plantings. Other species were used, but with varying success. Norway spruce was regularly planted and there were some large specimens in places such as Castle Menzies in Perthshire. However, its silvicultural requirements were as yet still poorly understood and it was regularly planted on unsuitable ground where it underperformed, inevitably making planters cautious in extending its use. European silver fir was also planted but on an even more limited basis.

By the end of the eighteenth century tree planting and ideas on the proper management of existing woods had gained sufficient status as a 'progressive' activity to be referred to by Burns who spent happy days in the company of the 4th Duke of Atholl at Blair Castle in 1787, exhorting him to plant still more trees. He made the Bruar Falls (see Plate 6.8) speak:

> Would, then, my noble master please
> To grant my highest wishes?
> He'll shade my banks wi' tow'ring trees,
> And bonie spreading bushes

It is hard to imagine that the Duke needed much urging. He duly obliged.

THE PLANT HUNTERS AND THE GREAT INTRODUCTIONS OF THE NINETEENTH CENTURY

By the start of the nineteenth century the vogue for forestry was well established, and the war against Napoleon both forced up the price of timber and concentrated the patriotic mind on the long-term need for shipbuilding timber. But perhaps there

PLATE 6.8 *The Falls of Bruar near Calvine, Perthshire, where Robert Burns successfully, as we can see, exhorted the Duke of Atholl in 1787 to 'shade my banks wi' tow'ring trees'. (Forestry Commission.)*

were even better trees which might grow in Scotland, if the world, rapidly opening up following the great voyages of discovery, was searched? Botanists and plant hunters invariably accompanied the explorers, looking for new plants with economic or aesthetic worth.

Scots, with their reputation as fine gardeners and hard workers and with a centre of excellence at the Royal Botanic Garden in Edinburgh, were to play a huge part in these plant-hunting expeditions, consequently introducing hundreds of new species of trees. Some of these were to have a major influence on forestry and the landscape of Scotland over the next two centuries. The story of their exploits is one of the most fascinating parts of Scotland's forest history.

Before the nineteenth century, introductions of new trees generally involved species from other parts of the British Isles, as with beech, or Europe, as with sycamore, larch, Norway spruce and European silver fir. Other species such as maritime pine and Cedar of Lebanon were tried with less success, though many continued to be used as specimen or ornamental trees. Trees from the eastern seaboard of North America were also planted extensively on a small scale from the early part of the eighteenth century, though ultimately this was to prove, by dint of the more Continental climate, much less suitable a source for introductions than the western seaboard. Weymouth pine and eastern hemlock, as well as a host of oak species and other broadleaves such as the tulip tree, were also tried but were to remain no more than fascinating specimen trees capable of growing in Scotland. The broadleaves in particular were highly valued for their autumn colours, but were not suitable as forest trees on any scale.

In the nineteenth century, the number and range of new tree introductions increased enormously, with North America, the Andes, the Himalayas, China, Japan, and Australia and New Zealand all proving happy hunting grounds for plant hunters looking for species that could grow in Scotland. The influence of conifer introductions was to be particularly important in widening the range of options open to foresters and landscape designers. They were all to take full advantage.

The first of the influential Scots plant hunters was Perthshire-born Archibald Menzies (1754–1842). Trained as a botanist and Royal Naval surgeon in Edinburgh, he was to travel around the world several times administering to sailors, which he did adeptly, and collecting plants, at which he was even more successful. Menzies, given his terms of reference by Sir Joseph Banks of Kew Gardens and Botany Bay fame, accompanied Captain George Vancouver around the world between 1791 and 1795 on a voyage of exploration and scientific investigation. He was famously to introduce the monkey puzzle tree from Chile after dining with the Spanish Governor of the region. On being served the seeds as a dessert, Menzies slipped some into his waist-coat pocket where they lay forgotten until his return to Britain in 1795. Monkey puzzles became a popular botanical curiosity and an outstanding feature of nineteenth-century gardens and designed landscapes.

A more economically significant legacy was to be his observations and plant specimen collections in the Pacific North West, a hitherto largely unexplored area. With a latitude and coastal climate similar to western Europe, the area of modern-day Oregon, Washington, British Columbia, and Alaska was subsequently to prove a treasure-trove of gigantic proportions, with an incredible range of new conifer species adapted to thrive in the British Isles, especially in Scotland. Captain Vancouver had used Douglas firs as replacement masts for his ships, highlighting the potential timber value. Menzies also collected Sitka spruce, ironically in the first instance to make spruce beer to ward off scurvy among the sailors. As a result of disputes with Vancouver, Menzies was unable to bring back seeds of the trees he observed there but his specimens and observations were to whet the appetite of plant hunters in Britain. Consequently in 1825 the Horticultural Society despatched a plant hunter (or 'scientific traveller' as they were more romantically called) to follow up his observations and send back collections to Britain. The man they chose for the task was a young Scots gardener called David Douglas (1799–1834) from Scone (see Plate 6.9), who had

PLATE 6.9 *David Douglas, 1799–1834, whose plant-hunting exploits in the Pacific North West were hugely to enrich the range of forest trees grown in Scotland – from a pencil drawing by his niece, Miss Atkinson, in 1829. (Reproduced by courtesy of the Royal Botanic Gardens, Kew.)*

come highly recommended by William Hooker, the Professor of Botany at Glasgow University. He was to prove an inspired choice. Ultimately he was to introduce over 240 new species of plants into the British Isles and his success in collecting and

PLATE 6.10 Portage on the Hoarfrost River *by Sir George Back, 1833, indicative of the conditions faced by plant hunters in the Pacific North West. (Reproduced by courtesy of the Glenbow Instititute.)*

PLATE 6.11 *The first Sitka spruce in Europe, grown from seed sent back by David Douglas, and planted at Keillour, between Perth and Crieff, in the early 1830s. Photographed in the mid-1860s. (Reproduced by kind permission of the Royal Scottish Forestry Society.)*

despatching viable seed of new trees and other plants was to change British gardens and landscapes enormously. Nevertheless it was for his tree introductions that he was to become best known, encouraging a frenzy for new trees and the establishment of specialist collections and arboreta across the land.

Among the many trees Douglas was to introduce were Douglas fir, Sitka spruce (see Plates 6.11 and 6.12), noble fir, grand fir, and Monterey or radiata pine. Such was his prolific output that he wrote to Hooker: 'You will begin to think that I manufacture pines at my pleasure.'

Sitka spruce and to a lesser extent Douglas fir were ultimately to form the backbone

PLATE 6.12 *A huge Sitka spruce felled in 1998, probably grown from Douglas's original seed and planted in the early 1830s, at Strathardle, Perthshire: not the popular conception of a Sitka! (Syd House.)*

of the current forestry industry in Scotland and the British Isles. Sitka spruce has been extensively planted to become the single most important forest tree in Britain, due to its dual ability to grow in relatively inhospitable conditions and produce quality timber. It now accounts for about two-thirds of all timber production in Britain. The stately Douglas fir is not so ubiquitous but continues to be a significant tree on more fertile sites, producing high-quality timber. Douglas's great contribution was to introduce these trees sufficiently early in the nineteenth century that, by the time of the founding of the Forestry Commission in 1919, foresters had had the benefit of ninety years of field observations to assess the real worth of these trees in forest conditions. This was to result in Sitka spruce contributing the bulk of the great expansion of tree planting in Scotland in the twentieth century, following the traditional pattern of trial as a specimen and cautious planting out on a small scale, before becoming the favoured tree of foresters and arguably (from an environmental perspective) ultimately being overplanted.

When Douglas's first tree seeds arrived back in Britain in 1826 and 1827 they caused great excitement. Seed was despatched to the members of the Society who had subscribed to Douglas's trip, including many Scottish estate owners. It was not uncommon for individual specimen trees grown from Douglas's collections to change hands for 15 guineas each, a substantial sum for the time and indicative of the frenzy for new specimen trees. Some of these original introductions of Douglas firs can still be seen in Scotland on Scone estate (see Plate 6.13), at Dawyck near Peebles and at

1 *The texture of a Scots pinewood. At Ledmore and Migdale in Sutherland, extensive regeneration and varied age structure provide an excellent habitat for wildlife. (Roger Warhurst/Woodland Trust Picture Library.)*

2 *Glen Tarff, Inverness-shire. Fragmented birch- and alderwoods cling to overgrazed slopes and eroded, treeless streamsides. (Rick Worrell.)*

3 *Glen Spean, Inverness-shire. Relict woods of birch, hazel, rowan and ash growing safe from sheep in a steep gully. (Rick Worrell.)*

4 *Scots pine on the tree-line at Beinn Eighe National Nature Reserve. To guess the age of such trees at the limit of their altitudinal range is very difficult, but some can be centuries old. (Laurie Campbell.)*

5 *The Wood of Skail, Strathnaver, Sutherland. Depicted on Timothy Pont's map of c. 1590 as a site where iron was made, it still grows where he depicted it, the vivid green beneath the birches delineating it from the moor beyond. (Anne-Marie Smout.)*

6 *Glen Strathfarrar, Ross and Cromarty. Much more varied in its character than most, this ancient wood of Caledonian pine and birch is relatively unchanged since the early eighteenth century, when the straths of the area were first described. (Neil Mackenzie.)*

7 *A stupendous newcomer. This Douglas fir grown at Dawyck Gardens near Peebles is from seed originally sent back in 1827 by David Douglas from the mouth of the Columbia River, USA. (David Knott.)*

8 *A veteran tree. The Birnam oak near Dunkeld, Perthshire, is associated with Shakespeare's 'Birnam Wood' in Macbeth, but was once more sinisterly called the 'Hanged-Men's Tree'. Notice the crucks that support the lateral branches, and the scale of the tree from the person at the foot. (Syd House.)*

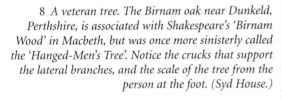

9 *Spring in Sutherland: Highland birchwoods in a sublime setting. (Anne-Marie Smout.)*

10 *Conifer plantings in Strathdon: surely the antithesis of natural woodland, but now a pattern of past practice. (Anne-Marie Smout.)*

11 *Felling an oak in Bailefuil Wood, Strathyre, for use in reconstructing the ancient roof of the Great Hall, Stirling Castle (late 1990s). Chainsaws were in universal use by the 1960s, and over the years became lighter, safer and more powerful, but it is still heavy work. (Forestry Commission.)*

12 *The march of mechanisation. This modern tree harvester is a modified civil-engineering excavator, working in a Perthshire wood. It fells the trees, removes the branches and then cuts the stem into billets for chipboard. (Forestry Commission.)*

THE TROSSACHS
BY EAST COAST ROUTE

13 *The romantic appeal of the Trossachs: a railway poster of the 1930s. (National Railway Museum: Science and Society Picture Library.)*

14 *Glen Trool, Kirkcudbrightshire: local people get involved in the management of their local oakwoods. (Forestry Commission.)*

Drumlanrig Castle, all places with exceptional collections and a strong history of pioneering tree planting.

Douglas fir was also to become highly prized for its landscape value and extensively planted as an ornamental specimen, particularly in policy woodlands around great houses. Indeed Douglas firs are now the tallest trees growing in the British Isles, with huge specimens over 200 feet tall on Loch Fyne-side in Argyll, the Hermitage in Perthshire and Reelig Glen near Inverness, testimony to its growth potential and

PLATE 6.13 *Noble fir growing in the pinetum at Scone Palace near Perth. Douglas began his career as a gardener at Scone, and the Pinetum also includes Douglas firs grown from seed sent back to his former employers. (Forestry Commission.)*

beauty. After Douglas's introduction there was a clamour to get more seed, especially as early observations on the success of the tree hinted at its forestry potential, but access to the Pacific North West for seed collection was not straightforward, so demand soon exceeded supply. Scone Estate was to profit from this demand, selling seed from the two 'Mother' and 'Father' Douglas firs, still standing today at Lynedoch on the banks of the River Almond. In the twenty years from 1850 to 1870, estate records at Scone suggest that some four million plants were subsequently raised from these trees, remnants of which are still scattered across the Tay Valley as well as in other localities such as the famous Sutherland's Grove in Barcaldine forest, Argyll. Douglas fir can grow to over 400 feet in its North American home and live for 800 years. Its potential in Scotland in terms of height and longevity is only just being realised (see Colour Plate 7), a living memorial to David Douglas, who has surely influenced modern Scottish gardens and landscapes more than any other single individual.

Douglas's field observations, made in the 1820s, on the trees growing in their natural habitat in the Pacific North-west show that he was aware of the ambitions of landowners who might wish to turn the unproductive bare uplands of Scotland into productive forest. He wrote in his *Journal* of Sitka spruce while surveying it on the banks of the Columbia river in Oregon:

> It may . . . become of equal if not greater in importance [than Douglas fir]. It possesses one great advantage by growing to a very large size . . . in apparently poor, thin damp soils . . . This unquestionably has great claims on our consideration as it would thrive in Britain where *P.sylvestris* [Scots pine] finds no shelter. It would become a useful and large tree . . . This if introduced would profitably clothe the bleak barren hilly parts of Scotland . . . besides improving the beauty of the country.

Douglas's observation was to prove prophetic in the twentieth century with the development of a major domestic timber industry based on the tree.

Even after Douglas's untimely death in Hawaii in 1834, the fashion for new conifer introductions from overseas (the 'conifer rage' as it was styled) increased enormously, driven by the usual dual purposes of profit and pleasure. Other Scots plant hunters and collectors such as William Drummond, William Murray and John Jeffrey were to botanise in the north-west of North America and send back dozens of other new species, such as lodgepole pine, western red cedar, Lawson's cypress and western hemlock, which in turn were eagerly planted by foresters and landowners when seed reached Scotland. Most became minor forest species fashionably popular for short periods. In the 1830s and 1840s other entrepreneurs such as the Edinburgh nurserymen Peter Lawson & Son (of 'Lawson's' cypress fame) were enthusiastically collecting and bringing back seed of other conifers such as stone pine and Corsican pine, the latter finding a niche for planting on the sand dunes of Tentsmuir and Culbin. In order to satisfy demand from potential purchasers, Lawson's sponsored their own North American collector, William Murray, such was the perceived commercial worth of obtaining new species and seed.

The story of the Oregon Association epitomised this drive to discover and introduce new trees. Originally founded as the 'Association for the Promotion of the
Arboriculture and Horticulture of Scotland', it had its origins in a 'meeting of gentlemen' held at the Royal Botanic Garden Edinburgh (RBGE) in 1849, with subscribers
including many of the great names and estates associated with Scottish forestry
such as Buccleuch, Roxburgh, Breadalbane, Lovat, Brahan, Fasque and Yester. One of
its main objectives was the continued botanical exploration of the west coast of North
America. Under the dynamic leadership of George Patton of Cairnies estate in
Perthshire, later to become Lord Justice Clerk of Scotland, the Association appointed
John Jeffrey, a gardener at the RBGE, as a plant collector. Jeffrey subsequently travelled to Hudson's Bay and retraced Douglas's path over the Rockies to the Pacific
North West. Between 1850 and 1854, before he mysteriously disappeared, Jeffrey sent
back several collections of seeds and other specimens to be distributed among the
Association's numerous subscribers. These included several new introductions such
as western hemlock and western red cedar as well as new collections of Douglas fir,
Sitka spruce, grand and noble firs and others. Many of the best specimens of these
trees can still be seen growing in the policies and grounds of the subscribing estates
scattered across Scotland.

One of the most popular and eagerly awaited introductions was the giant redwood
(*Sequoiadendron giganteum*) from the Sierra Nevada in California. Because of the
writings and work of another Scot, John Muir, who had emigrated as a young boy to
the USA, ultimately becoming a major influence in the founding of the National
Parks movement, the giant redwood has always had a special interest to Scots. The
first introduction of these trees into Europe was in Perthshire in 1854 from seed sent
back from California. Many of these original trees still survive, including one tree at
Cluny Gardens, near Aberfeldy, measured recently as the widest conifer in the British
Isles with a girth of over 11 metres. The giant redwood was to prove hardy throughout
Scotland and subsequently to become one of the most popular trees planted in policy
woods and pleasure gardens.

While trees from the Pacific North West were to prove the hardiest and most
productive of the nineteenth-century introductions, the other great source of new
conifers was Asia – the Himalayas, western China and Japan in particular. Despite a
huge variety of potential candidates, species introduced from Asia were generally to
prove less economically useful. The exception was the Japanese larch, which was
introduced to Scotland from the slopes of Mount Fuji in 1883. Japanese larch inhab-
its a small niche in Scottish forestry, partly in its own right as a forest tree resistant
to the canker which affects European larch, but more importantly in hybridising with
the European larch to produce the Dunkeld or hybrid larch (*Larix x eurolepis*) – first
recognised in the early 1900s in seedlings growing close to the 'Parent' larch in
Dunkeld and now the preferred larch planted in Scotland – Scotland's own unique
contribution to forest trees.

Other tree introductions such the eucalypts from Australia, southern beeches
(*Nothofagus*) from Chile, and the cedars – the Atlantic cedar from the Atlas mountains of Morocco, and the deodar cedar from the Himalayas – were introduced in

this period. All are hardy and have gone through fashionable periods of planting, ultimately to remain as popular amenity trees of landscape and specialist interest.

The plant collectors' efforts led not only to the establishment of a number of superb specialist collections of trees such as at Benmore Gardens in Argyll and Dawyck near Peebles, but also to a fashion in landscape design utilising these new trees to enhance the surroundings of the houses of the rich and wealthy across Scotland. New country houses were invariably sheltered by Douglas firs, silver firs (*Abies* species), pines, spruces and cypresses, while specimen monkey puzzles and giant redwoods crowned this arboricultural enrichment. The attraction of these trees, blended in with the more traditional broadleaves, was the fact that they were evergreen and provided year-round colour in a country where most of the native broadleaves are bare of foliage for nearly six months of the year. The grounds of Dunkeld House, now a hotel, features an 'American' garden of many of the introduced conifers now grown into gigantic specimen trees. It is one of the best examples of the impact of this approach on landscape planning and design. This use of ornamental planting was also a feature of urban planting in public parks such as Kelvingrove Park in Glasgow and even in public cemeteries such as Tomnahurich Cemetery in Inverness.

For wealthy landowners the new trees were used to enrich an already well-wooded landscape, sometimes purely for ornamental reasons and sometimes for sporting purposes, or, as already described, to satisfy their urge, often in competition with their neighbours and peers, to collect and display their prizes like some sort of huge gardening show. In recent years the work of the Tree Register of the British Isles, the Garden History Society and the *Inventory of Gardens and Designed Landscapes in Scotland* has gone a long way to build on the work of the late Alan Mitchell to catalogue and record the content of these collections and the specimen trees contained within them. The number of champion trees, especially conifers, and original introductions, is testimony to the value and breadth of these collections, the efforts of generations of collectors and growers and the suitability of the sites and climatic conditions.

Murthly Castle in Perthshire is one such magnificent example with sweeping avenues of oak and cedar trees lining the approach from the west and ancient limes and giant redwoods from the east. Below the plateau on which stands the castle, there is probably the finest example of the use of these introduced conifers. Curving avenues of giant Douglas firs dating back to the 1840s lead to groves of other conifers, including several original introductions such as western hemlock and Serbian spruce.

Ardkinglas estate on the shores of Loch Fyne in Argyll is another excellent example (see Plate 6.14). Building on a history of woodland planting throughout the eighteenth century and continuing the development of the greater Ardkinglas policies, a pinetum was established in 1875 to take advantage of the trend for planting the new conifers. Conditions for these new introductions proved to be excellent, with Argyll rivalling Perthshire as Scotland's premier location for growing conifers, to the extent that Douglas firs and grand firs dating from this time from both localities challenge as the tallest trees in the British Isles.

Visitors to many of the great houses of Scotland, such as Dalkeith Palace, Mellerstain, Drumlanrig, Scone Palace with its extensive pinetum, Blair Castle with the

PLATE 6.14 *Silver fir (Abies alba) in Ardkinglas Woodland Garden, Argyll, around 1870 and once described as 'the mightiest conifer in Europe'. Perhaps the two shooters are waiting to ambush roe deer. (Reproduced by kind permission of the Royal Scottish Forestry Society.)*

magnificent collection of giant conifers at Diana's Grove and a thousand hectares of other woodland surrounding it, and Inveraray Castle in Argyll, can observe the impact of these impressive tree collections and their use in landscape enhancement on a large scale. Dawyck and Benmore Gardens have been adopted as outliers of the RBGE, such is the value of their tree and other plant collections.

Forestry in the Victorian and Edwardian Age

In the period up to 1850, planting continued apace, matching that of the previous century. A census of woodlands in 1812 had indicated some 914,000 acres of woods in Scotland, of which around 45 per cent were recorded as planted, the remainder being, presumably, semi-natural woodland remnants. This would suggest a rate of planting over the previous 100 years of around 4,000 to 4,500 acres of new planting per annum, a remarkable achievement given the modest beginnings. A further census in 1845 suggested that the area of planted woodland had increased to around 595,000 acres, a planting rate of between 5,000 and 6,000 acres per annum of new woods (in fact it was probably greater than this as some of the earlier plantations would have been cut down and not replanted as there was no requirement to do so) with the greatest area of plantations in the counties of Inverness (87,000 acres), Aberdeen (86,000 acres) and Perthshire (64,000 acres). In Inverness-shire the main planting had been undertaken in Strathspey on Seafield estate in particular, where the remnant native pinewoods had been extensively added to by planting with Scots pine of local native origin as well larch, and on Lovat estate as a result of the positive approach undertaken by the late Commissioners of the Forfeited Estates. Scots foresters had become leaders in the silviculture of conifer plantation forestry.

The development of such a large area of productive woodland had the potential to establish a strong Scottish timber industry as had been forecast by planters such as the 4th Duke of Atholl. For much of the first half of the nineteenth century, forestry as an economic activity prospered. With timber prices high and labour costs low, the outlook was good. Such optimism unfortunately was sadly misplaced. Cheap timber imports from the seemingly limitless supplies of the British Empire, Russia and Scandinavia, the abolition of duties on timber imports in 1866, and the replacement of timber with iron for shipbuilding, all were to change the financial climate for the worse for Scottish timber growers in the latter half of the century. In terms of free trade to encourage industrial development and support the burgeoning population, these moves were understandable but the result was that home-grown timber simply could not compete in the face of such competition, a major disincentive to active woodland creation and continued management. The government of the day, having already accepted cheap food imports as necessary to supply the increasingly urban population, was not interested in having any forestry policy when timber could be cheaply imported from overseas.

Land use was changing too, especially in the Highlands, where the management of open hill land for red deer was becoming increasingly important, heavily dependent on sporting lairds who inevitably had little interest in a long-term commitment to woodlands except as shelter for game and deer. In the face of such disinterest, forestry as an economic activity declined in comparison with the enthusiasm of the earlier age. A further estimate of woodland cover in Scotland in 1872 suggested it had fallen to 734,500 acres as woods were felled and not replanted.

The picture was not entirely gloomy, however. There continued to be a keen interest in amenity and landscape planting as evidenced by the large-scale use of conifers in landscape design around the great houses across the land, many of which were rebuilt, altered and generally 'improved' by the influx of wealth from the Empire and trade. The pattern of land ownership, too, was changing as the new class of wealthy industrialists and merchants purchased Highland estates and wished to emulate (and surpass in opulence and style) the lifestyle of the landed gentry. It became the norm to establish policy woodlands around Victorian houses of any consequence. In the countryside the spirit of 'improvement' and making 'wasteland' productive also remained very strong.

Ideas on how to practise forestry were also developing strongly, much influenced by German 'scientific forestry'. Even if forestry in Scotland did not offer great challenges, the management of forests across the British Empire certainly did. All of this encouraged the establishment of schools of forestry at the Universities of Edinburgh, where Dr William Somerville was appointed lecturer in 1889, and Aberdeen, where lecturing began in 1907, the better to disseminate sound practice to students and encourage research on a proper scientific basis. In 1854 the Scottish Arboricultural Society, to which 'Royal' was added in 1887, was founded. Now known as the Royal Scottish Forestry Society, it is the oldest forestry society in the English-speaking world.

In terms of establishing new woodlands, the introduction of new conifers continued to be the most important development of the age, along with a growing professionalism in the practice of forestry. The basics of good silviculture when establishing and managing woodlands were clearly recognised and set down – clear objectives of management, proper species selection for the site, the correct seed-source for planting or regeneration, adequate site amelioration such as cultivation, fertilising and weed control, appropriate management of the growing crop by cleaning out competing species, thinning to allow favoured trees more space to grow, and protection against pests, generally of the four-footed kind, and diseases. In 1847 *The Forester* was published by James Brown, forester on the Arniston estate, Midlothian. It became a popular book and marked a rise in the status and professionalism of the estate forester.

Conifers were by now much the preferred types of trees for planting outside the better ground in the Lowlands, with Scots pine, larch and Norway spruce the dominant species. The choices available to foresters were changing, however. In 1834 the first recorded planting of Sitka spruce took place at Keillour in Perthshire (where it is still grown today with at least three rotations harvested to date) and thereafter there was an explosion of interest in new species, particularly Douglas fir.

Despite the lack of clear commercial viability and the absence of any state support for the business of forestry there was still some enthusiasm for establishing new woods, partly because of the traditional enthusiasm of landowners allied to a strong interest in game management, which new woodlands would enhance. For the period from the 1860s to 1913 the rate of new woodland establishment was around 4,000 acres each year. Although larch and Scots pine remained the main species where timber production was the overriding objective, significant use of the new species was made in the numerous new woodlands planted in the period leading up to the

While the overseas connections must have been a good selling point for the exhibition, the Society's real interests were a lot closer to home. It achieved its educational objective very quickly (when the University of Edinburgh established a lectureship in forestry in 1889) but then went on to develop an ambitious campaign for 'national demonstration forests' and for a government Board of Forestry – here

PLATE 7.2 *Simon, 16th Lord Lovat, whose energy and vision did more than anyone to create the impetus for a Forestry Commission: he was keenly interested in land improvement, and stemming the drift of population from the Highlands. (Forestry Commission.)*

were the beginnings of a national forest policy. Between 1900 and 1914 it contrived a whole series of inquiries into forestry and, along with other bodies, lobbied persistently in Westminster and Edinburgh. But its efforts got it nowhere – the governments of the day seemed impervious to reason. The Society found it all very frustrating.

That was until the outbreak of war, fertile ground at last for the Society's ideas. In 1916 four of its members were drafted on to the Acland Committee (after its Chairman, Francis Acland MP), which was set up by the War Cabinet to advise on 'the best means of conserving and developing the woodland and forestry resources of the United Kingdom, having regard to the experience gained during the war'. We shall hear later about Lord Lovat (see Plate 7.2) and Sir John Stirling Maxwell of Pollock, and about the Committee's Secretary, Roy Robinson (see Plate 7.3).

PLATE 7.3 *Roy Robinson (Lord Robinson), who became the Forestry Commission's first Technical Commissioner in 1919, and later its Chairman, until he died in 1953. (Image courtesy of the Forest Research Library.)*

PLATE 7.10 *Drifter on the stocks at Findochty, Banffshire: probably 1920s or 1930s. Scots- grown oak and larch were the main timbers used for boat building until steel took over in the 1960s and 1970s. (Courtesy of the Scottish Fisheries Museum.)*

Cairngorms (1948) and the Queen Elizabeth Forest Park (1953), so named to mark the coronation of the Queen, and which, at the start of the new century, will play a central role in Scotland's first National Park, Loch Lomond and the Trossachs.

We should pause briefly to reflect on Robinson's extraordinary career in the Forestry Commission, because he set the tone for its activities between 1919 until long after his death in 1952. Robinson was an Australian Rhodes scholar who had started working in the Board of Agriculture and was Secretary to the Acland Committee, before becoming the Commission's first 'Technical Commissioner' and Accounting Officer. His initiative on Forest Parks (which was the beginning of the Commission's considerable future investment in recreational facilities for the general public) seems to be the clue to his real interests, which were more about land use than about wood-using industry – something of a paradox in view of how the Commission's priorities would later be seen.

In 1943, by then as Chairman, he was heavily engaged in planning post-war forestry and it was obvious that what was decided would shape the pattern of woodland activity for many years to come. We can imagine that, behind closed doors in Whitehall, this was the first in-fighting between the farmers, the foresters and the newly-emerged camp urging the creation of National Parks. According to Stirling Maxwell's obituary of him in *Scottish Forestry* in 1952, Robinson needed all his diplomatic skills in the negotiations which were 'prolonged and acrimonious though they had a smooth passage through Parliament'. Of course Robinson was an old hand in the politics of Whitehall and would have been well prepared for all the arguments, even if the outcome might not have been quite what he wanted.

THE SECOND WORLD WAR AND ITS AFTERMATH

We can only wonder what difference it might have made to forestry policy in 1919 had it been known that another strategic emergency would occur so soon. The woods planted in the inter-war years for the very purpose of providing for such an event were simply too young at the outbreak of the Second World War. So, for a second time, the nation had to fall back on the woods of the private landowners. By the end of the war, some 230,000 acres (92,000 hectares) of them had been felled in Scotland.

At least the authorities were better prepared than they had been at the outbreak of the First World War, and the organisational and operational tasks were more familiar. Tree harvesting was still being done by men and horses, even if modern methods (as they then were) using 'crawlers' and wheeled tractors for timber extraction had just begun to creep in. Here, for example, are A. & R. Brownlie's recollections of their work on the Tyninghame estate in East Lothian in 1941:

> Brownlie's were allocated more than 650,000 cubic feet of standing timber by the Timber Control to be felled for the war effort. The remaining timber on the estate was allocated to James Jones of Larbert. The last tree to be felled by Brownlie's in the Binning Wood on the estate was a magnificent oak more than 250 years old. Brought down by axe and handsaw, the wood was extracted by crawler and then taken . . . to the nearby railway station. Horses were not replaced by mechanised vehicles during the war . . . Brownlie's for example still had 80 pairs of horses working in the woods at the time. At Gala Hill at Galashiels, beautiful beech trees were felled, converted and the boards stacked to season. The woodcutters cut limbs from the top of the trees to prevent them splitting as they hit the ground, hauling themselves up the trunks using climbing irons. The beech went to Mosquito aircraft factories and bobbin factories. At Earlston, the saw-mill turned out timber for targets, buffer blocks, and keel blocks for the Admiralty.

As in the First World War, a 'Home Grown Timber Department' was set up to work timber and operate sawmills, the majority of whom were civvy-street recruits who had to learn their new forest skills in just a few weeks. A Women's Timber Corps, formally inaugurated in Scotland in May 1942 at the very peak of the wartime labour demand, worked alongside the Department's employees all over the country (see Plate 7.11). Italian and German prisoners-of-war were assigned to forestry work.

A new generation of forestry workers were recruited in several Commonwealth countries. The Canadians and 'Newfies' returned, and there were foresters from Australia and New Zealand. In one of the more bizarre episodes of the war a unit of

PLATE 7.11 *Women to the rescue. Forestry was a male-dominated industry, but in both world wars women helped with tree-felling and other forest work. In 1942, a Scottish Women's Timber Corps was formally set up, and its members worked all over Scotland: this picture was taken on the Bowmont estate near Kelso. (© The Trustees of the National Museums of Scotland.)*

500 men from British Honduras were employed in the woods on the Dunrobin estate near Golspie, suffering terribly from illness and cold.

This time the search for timber really went into the nooks and crannies of the

PLATE 7.12 *During the Second World War, the supply of wood for the mines was critical. This winter scene is from West Linton, outside Edinburgh.* (© The Scotsman *and National Museums of Scotland.*)

countryside – places missed in the First World War or woods that were too young then or too difficult to get at. Consider, for instance, felling trees on Inishail Island in the middle of Loch Awe. In *Timber!*, a delightful collection of rose-tinted reminiscences by members of the Women's Timber Corps, Margaret Grant remembers working there:

Rafting was done in two particular places. One, Inishail, a little island in the middle of Loch Awe, that we called the 'The Green Island', though in Gaelic it means the stately, charming isle, was heavily wooded with tall larches and spruce. The bracken up from the shore grew shoulder high and in the centre were the ruins of an old monastery. At one end, sheltered by trees and overgrown with roses and briar, was a tiny place with old, old stones, like the ones in Iona. We worked on this island, cutting wood, so there should have been a great deal of noise. I used to stop and listen for the ring of axes and muffled crashes, the clanging of chains and the clumping of horses on the brown and peaty drag road, the crackling and hissing of fires burning and the hammering of iron 'dogs' into the rafts; sounds of beasts and humans working together though it never sounded like another work-a-day wood. The sound seemed to be enveloped and taken up as if there were a great blanket of silence over everything. The quietness was a wonderful experience.

PLATE 7.13 *A young miner trained in setting a 'tree' or prop in the 1950s. The coal mines pro-vided a huge outlet for the smaller logs produced in thinnings, but by the 1960s the use of timber was being phased out as small pits closed and steel jacks came into use. (Courtesy of the Scottish Mining Museum Trust.)*

We also rafted from Port Sonachan, a beautiful place where there is a hill on which we had cut all the timber, right across to Kilchrenan on the other side. I loved rafting: standing in the water in waders with a big hammer through your belt, the chain from the raft over your shoulder, with 'dogs', rings with spikes on them, and the men coming down with the horses.

The colourful story of the Canadians in the Highlands has been researched by William Wonders for his book *The Sawdust Fusiliers*. A roll-call of the landed estates that hosted the Canadians would do justice to any present-day guide to forestry in the Highlands – Atholl, Ballogie, Balnagown, Balmoral, Cawdor, Darnaway, Glentanar, Lovat, Moray and Novar, to name just some of them. They brought with them the most up-to-date logging equipment then available in Canada: caterpillar tractors for extracting the timber, fast diesel-powered saw benches, winches for high lead logging and bulldozers for road-building. It was a lesson in heavyweight logging – as much an eye-opener to the home squads as it was an adventure to the young Canadians. And for some of them it was an opportunity to search out their family roots:

The Highlands are showing . . . an increasing number of bare patches on their hillsides. But sentiment must give way in time of war . . . This company all hail from the neighbourhood of Kirkland Lake, Ontario, and includes French Canadians, Scots Canadians and Canadian Indians. They have two pipers – Alexander Shanks and Maurice MacDonald . . . Piper MacDonald told me that not long ago he visited Skye, where his ancestors came from, and although he found none of his people there he evidently treasured the memory of his visit. Canadians are Canadians first and foremost; but they never fail to keep a place in their hearts for the ground where their parents, or grandparents or great grandparents were born.

But of course the real story was the destruction in the woods. Frank Fraser Darling (later Sir Frank), naturalist and author, described the scene in a paper to the British Association as he saw it in 1949:

Our land is so devastated that we might as well have been in the battlefield. See the very windbreaks taken from the roadsides, see the wreck of Glenfeshie, the Rothiemurchus that is no more, the shearing of the Sutherland woods, the removal of shelter from the wind-ridden West and the immense landslips on the road from Tomdoun to Loch Hourn. That is what deforestation has meant in our time.

The Second World War had a profound effect on every aspect of country life. The landowners and farmers entered a new world of intensification, subsidies and increasing controls. We shall see later how the government increased afforestation but it also determinedly set about restoring the wartime damage to the landowners' woods, bringing them back into productive use and even hinting at compulsory state purchase if they were not taken in hand. This was a complete reversal of the laissez-faire attitude that had prevailed after the First World War, which in 1940 provoked Sir John Stirling Maxwell (as reported in *Scottish Forestry*) to urge the government 'to bring to its task far more energy and imagination than was shown after the first war'.

Sir John and his beloved RSFS, which he considered 'had always done the hard thinking on forestry in Scotland', can take some credit for the government's change of heart after 1945. It was, together with its counterpart society in England and Wales, the only representative body seriously lobbying for the care of woods at that time, and we can surmise that he, more than most, knew how to make things happen. He had, after all, been a Forestry Commissioner from the beginning and its Chairman between 1930 and 1932. At about the same time he was involved in the creation of the National Trust for Scotland. Furthermore, he could make his point with all the confidence of being a woodland owner and silviculturist himself, much taken with the practical challenges of growing trees on the high-elevation shores of Loch Ossian on his estate at Corrour.

The Society's promptings did not go unheard. The 'Dedication Scheme' – launched by the post-war government in 1947 – encouraged landowners to commit their woods to forestry in perpetuity in exchange for financial help with restoration and replanting. It was greeted with suspicion at first but eventually caught on and became the mainstay of post-war forestry. Landowners quietly got on with the job of rebuilding their woodlands encouraged by a political consensus on forestry policy.

Dedication came in all shapes and sizes. Most woodland estates were a few tens or hundreds of hectares; only a very few were over a thousand. The late Duke of Atholl, speaking at the Birnam Conference in 1988, was reflecting the views of many landowners when he applauded its achievements:

> The Dedication Scheme imposed real burdens on the landowners...and was recorded on countless land titles throughout the country [but] it was a far-sighted and successful scheme in that it formed the base for the great expansion of private forestry which has taken place in Scotland since the war. Along with the fiscal arrangements applied to forestry, much inflated by sky-high taxation, planting became an excellent investment and outside capital poured into the industry. It was also the spur to the investment which has recently taken place in the wood-based industry and which has altered the whole outlook for Scottish forestry.

The Dedication Scheme had much to commend it. It meant a long-term commitment to the management of a renewable natural resource, a remarkable turnaround on the attitudes of the first forty years of the century. However, it was a creature of its time, slanted to the government's main purpose of wood production. We shall see how, in the late 1980s, this policy changed. By then, the Thatcher government had swept away Dedication in a political climate of market reform. It was closed to new entrants in 1981, ostensibly to reduce cost and 'red tape' and was replaced by a regime of short-term grants.

It is opportune here in our story, before going on to the post-war phase of afforestation, to say a few words on the academic background of the foresters in Scotland; no account of twentieth-century woods would be complete without a mention of forestry teaching. It was delivered by several institutions at various levels but only Edinburgh University can claim to have gone the full course of the century. It is perhaps most fitting to mention Mark Anderson, Professor of Forestry at Edinburgh University from 1951 until he died in 1961, because in one particular way his story provides a link to the present day.

Anderson was a man of learning, a prolific note-taker, author and translator of specialist forestry books in Russian and German. He is said to have taught himself Russian in the trenches of the First World War. He is probably best known for his monumental *History of Scottish Forestry*, which was put together by Charles Taylor after Anderson's death. Taylor saw Anderson's forestry philosophy as a desire to 'study nature, follow her if you can, but guide her where need be and record what is done and achieved'. His students experienced the reality of this in his advocacy of the 'group selection system' and the 'check' method (of control), modelled on the mixed irregular forests of the Swiss Jura, which he emulated in the so-called 'Anderson plots', which can still be seen today in the woods at Glentress. What did not figure in his teaching, but which he energetically promoted outside the lecture hall, was the case for a separate forestry policy for Scotland. To his students he could seem dour at times – a man lost in his books – but had he been alive in 1999, he would have looked up and smiled when political responsibility for forestry passed from Westminster to the Scottish Parliament.

The Dedication Scheme had been just one side of a two-pronged post-war forestry policy. The other side was to be an increase in the Forestry Commission's afforestation programme. But there was a subtle and significant change in comparison with the pre-war policy. Absolute priority was to be given to agriculture, using marginal land wherever it could be converted to use and the climate allowed it. The big ploughs that were being developed for forestry could now be used to expand food production. The government announced that afforestation would be confined to the uplands where it would not use up productive ground.

The significance of this is not quite as it might seem. Upland conditions in Scotland come down almost to sea level, so in one sense the new policy made no difference. But its effect was almost to rule out further new planting in England, so concentrating the whole of the national programme in Scotland and Wales. The scale of the task envisaged had many implications, some of course of a very practical kind. How to get people to work there? Now the talk was of forest villages (see Plate 7.14).

'Moving to a new life, in a new street, in a new village – in a new forest' was how the Scottish Daily Express reported on a newly married couple moving into their first house in the Minnoch village at Glentrool on 25 October 1955. It was one of around forty forest villages built by the Forestry Commission and the Scottish Special Housing Association in the boom of enthusiasm after the war, in what were expected to be the great centres of the burgeoning afforestation programme, particularly in the west of Scotland and the Great Glen.

There was a feeling of optimism about the swelling numbers. The forest village of Dalavich on Lochaweside was started in 1950 and finished in 1955, when it comprised a total of forty-seven houses, a school, a post office and a community centre. There were then 318 people resident in the Inverliever area as compared with only 55 when the Office of Woods started its activities there in 1908. But the policy of building new villages in the forest, rather than attaching them to established villages, was often criticised – there was little opportunity for people to broaden their social contacts; car ownership was unusual and public transport was poor. Almost fifty years on, the villages are now no longer mainly occupied by forestry workers. But if, in retrospect, they were not the success their planners had hoped for, they nevertheless continue today to provide rural housing and spending power.

Ironically the Forestry Commission's workforce in Scotland reached its very peak in 1954, boosted by the push of recruitment for its villages. It reached 5,071 compared to about 2,000 at the end of the century. Forestry mechanisation was the main reason for the decline. By the 1950s James Cuthbertson of Biggar was making his mark. His was one of a number of small engineering firms that brought their skills to the cause of the foresters. Cuthbertson developed ploughs to tackle the peaty slopes of Dumfriesshire

PLATE 7.14 *The forest village of Glenbranter, Argyll, newly built for the Forestry Commission in the 1950s. Such housing has been largely sold off since, when the industry failed to produce the anticipated level of employment. (The Scotsman and the Forestry Commission.)*

and much of the landscape beyond, eventually commercialising his machinery and selling it all over the world (see Plate 7.17).

Cuthbertson heralded an era of mechanisation which enabled foresters to grow trees almost anywhere – draining and ploughing methods for every type of soil, fertilising from the air to remedy nutrient deficiencies in the upland soils, and above all tree planting with Sitka spruce, widely planted before, but now becoming an absolute *sine qua non* for its remarkable tolerance of poor soil and exposure to wind. By that time it was gaining some favour with the cautious wood merchants as well.

THE GREAT SURGE IN PLANTING

By 1960 Scotland's woodland cover was only 6 per cent, not a lot more than it had been at the beginning of the century. So it seemed that the afforestation so far had done little more than substitute for the abandonment of so many of the old woods after the First World War. Now, however, the scene was set for a surge in planting and for some dramatic changes in the landscape, until by the end of the century about 17 per cent of Scotland was forested.

It followed from proposals made by Sir Solly Zuckerman in his Report to government in 1957, which sidelined the *strategic* objective that had been the main purpose of the afforestation programme since its inception in 1919, and proposed instead that forestry should have economic and social aims. It made the foresters concentrate on improving the financial returns from their tree planting and, as we shall see, on developing the wood-using industry. John Davies, writing in *Scottish Forestry* in 1971 about his 'Argyll Tree Farming School', saw it thus: 'our skill then as foresters does not lie in copying nature, but in bending natural forces to our will. The successful forester is the one who in spite of natural forces and difficulties produces the most valuable crop most economically.' It was, in the uplands, a harsh and, some would say, harmful planting philosophy, which, even in the 1960s, was beginning to draw criticism from the general public. Part of the problem was that those geometric and bland looks invited complaint as the motorcar brought increasing numbers of visitors to rural Scotland.

It would be a long time before the amenities of forestry in the uplands were fully addressed, even if, by the 1960s, the Forestry Commission had become fully alive to the problem. In 1963 it turned to Sylvia Crowe (later Dame Sylvia) to advise them. She was a past President of the Institute of Landscape Architects and an adviser to the Central Electricity Generating Board. In her book *The Landscape of Power* (1958) she wrote with passion about artificial structures in the landscape:

PLATE 7.15 *Glentrool Forest Park, opened in 1947, was the second of Scotland's Forest Parks designated by the Forestry Commission to encourage recreational use of the forests. This photograph was taken in the 1960s. (Forestry Commission.)*

There can be no doubt that men have not learned to reconcile their activities with the ecology of the earth, either visually or organically, and until they have done so it behoves them to guard the background of tried and balanced nature, with all the skill and power which the new machines should give them, directing a small proportion of their new wealth to the well-being and appearance of the land.

Crowe was raising some fundamental questions about the use of land which bear on what we now call sustainable development. By the 1990s sustainability had become a political ideal brought into everyday parlance by the Rio 'Earth Summit' in 1992 and epitomised on a global scale by issues like climate change. At home, in forestry, 'sustainable forest management' preoccupied the minds of foresters and ecologists – how to define it and how to implement it. Much was done in the last two decades of the century to improve forest practices and to remedy the deficiencies of the early afforestation. We take such concern for granted today, but Crowe was writing in 1958. It was, for example, only seven years earlier (in the Forestry Act of 1951) that the law, for the very first time, required the owners of woods to replace them after felling.

Under Crowe's guidance, forest planning became more analytical, more of a reflection of natural shape and form, and less stereotyped. She preached that the artificial shapes of the forest needed to reflect the changing colours and textures of the seasons (by increasing the planting of the deciduous larch, for instance) and to take account of the growth of the woods over time. Her casework with the Commission was not just about afforestation but about the shapes and sizes of tree-felling areas. By the 1960s the timber trade had started a progressive change-over from the mixed woods of the private landowners to the coniferous afforestation resources of the Forestry Commission. And they had begun the planning for one great wood-using enterprise which was ultimately unsuccessful but which improved the whole climate of opinion in the wood-using industry that the foresters courted.

This was the Scottish Pulp and Paper Mills at Fort William. Lord Polwarth, Chairman of the Scottish Council, cut the first sod there in July 1963 in the confident expectation that 'for the first time in 250 years, we will see repopulation and not depopulation in the Highlands'. It was welcomed in similar terms by both houses of Parliament and, uniquely, a special Act of Parliament was passed to enable the government to grant it a loan. To the woodland owners its creation had seemed a symbolic event which lifted them 'out of a rather dismal pit prop mentality and put a different complexion on their forests'. Wiggins Teape (the parent company) were at the outset delighted with their investment.

Whereas before, the use of axes and handsaws for tree felling were the norm, the woods now saw the introduction of the labour-saving petrol-driven chainsaw. The extraction of timber, which was taken over by tractors and winches after the war, was still being done by horses in the hill forests until, for Fort William, they were retired in favour of 'cable cranes' and skidders. What ten men would have done in the 1950s, one man could now do in the rather noisier woods of the 1960s and 1970s.

People said that the pulp mill saved the West Highland Railway from the attentions of Dr Beeching, a reprieve of course welcomed, except by a few nervous landowners with woods on the line. According to the *Journal of the Forestry Commission* in 1949, there was an 'inconvenient gradient just outside Fort William, which necessitates heavy stoking by the firemen and much puffing by the engine. At times . . . this scene resembles the fifth of November.' Sparks from the steam engines regularly set fire to the Moor of Rannoch and the woods around!

But the effect on landowners generally was more positive, and the arrival of the pulp mill sparked off a new kind of landownership, exemplified by the forest of Eskdalemuir in the high hills of eastern Dumfriesshire between the Esk and the Ettrick. In 1965, the Economic Forestry Group acquired a large area there for afforestation by private individuals. It was an area of sheep farms in decline, with stocking levels down to less than one ewe for every hectare. One by one the farms were sold for forestry and planted up with trees, eventually creating a forest of around 22,000 hectares in seventy distinct ownerships.

The government's plan for post-war afforestation seems to have completely over-looked a role for private forestry, and so it might have remained if Kenneth Rankin, a city accountant, had not made the connection between the long-term sureties of Dedication and the benign way that forestry was taxed. It meant that people's tax liabilities could be diverted into woodland creation. It was a perfectly legitimate way of saving tax, although often represented by its critics as tax avoidance. Rankin saw it as a way of converting income into a natural capital asset, as well as laying the foundations of a new industry in which he was a great believer.

It was also a business opportunity. Rankin had an infectious enthusiasm, and takers lined up, so much so that he set up his own afforestation company, the Economic Forestry Group, and inspired the creation of other forest management companies, of which Tillhill, Fountain Forestry and Scottish Woodlands became particularly well known. Between the 1960s and the 1980s, the companies together planted many thousands of hectares of coniferous forest in Scotland, particularly in Galloway and the west.

However, company afforestation eventually got into trouble with serious consequences for private forestry as a whole, as we shall see. There were no planning constraints or consultation on change of land use when the companies bought land for afforestation and their 'go anywhere' style seemed careless of other interests. This brought it into conflict with other land users, particularly with nature-conservation interests, and in the 1980s led to a number of high-profile rows about planting schemes.

The Wildlife and Countryside Act of 1981 contained provisions whereby, within Sites of Special Scientific Interest, landowners could claim financial compensation if their operations were not permitted for wildlife conservation reasons. When a forestry company acquired and proposed afforestation at Creag Meagaidh, a hill massif above Loch Laggan, it seemed to the public like a piece of opportunism, prompted by little more than the chance to get a pay-out from the taxpayer. The dispute was eventually resolved when Creag Meagaidh was sold to the Nature Conservancy Council, who gradually turned it into a model demonstration site for the natural regeneration of native broadleaf woodland.

PLATE 7.16 *A crawler tractor pulls a tine plough through hard mineral soil, near Bonchester Bridge, Roxburghshire, 1976. Cultivation here was complete, not in strips. (David Foot.)*

And then there was the Flow Country, a flat wet basin in Caithness and Sutherland and a uniquely-special habitat for moorland birds, brought to the public's notice by the Royal Society for the Protection of Birds (RSPB). They are called 'the flows' after the shimmering pools and ridges of sphagnum moss which, from above, seem to flow across the landscape. In 1983 the Society discovered that Fountain Forestry were buying land there for afforestation that would completely destroy its special nature interest. 'A disaster waiting to happen somewhere' was how Mark Avery and Rod Leslie saw it, as two 'insiders', in their book *Birds and Forestry*.

The RSPB's campaign turned over a stone on some uncomfortable truths. First, if the area was so important for wildlife, why did it not already have statutory protection? Why was there no consultation on these forestry schemes? And what was the real case for forestry anyway? People argued that the land was too poor for trees and too remote from markets. It was a dispute that seriously discomforted the public authorities responsible for land use and embarrassed the government. It went on for all of five years in an atmosphere of increasing acrimony.

Then, in the 1988 Budget, Chancellor of Exchequer Nigel Lawson changed the tax rules that had underpinned company forestry. It was a big blow to the forestry companies and the event that is often pictured as the final curtain on large-scale upland conifer afforestation. The reality was more complicated. Perhaps the productivist

PLATE 7.17 *The Cuthbertson ploughs, named after James Cuthbertson of Biggar, transformed the possibility for planting on high and peaty ground. Nearly all post-war afforestation took place on land ploughed to improve drainage and prepare for trees. (Image courtesy of the Forest Research Library.)*

ethos that had prevailed since the Second World War was running out of steam anyway. In the previous year, the government had introduced a pilot Farm Woodland Scheme – later developed into the Farm Woodland Premium Scheme – which offered farmers annual payments to encourage them to convert productive agricultural land to woodland. The new approach was underlined by Lord Strathclyde, the Minister of State at the Scottish Office, in 1991: 'what is happening is that forestry is emerging from its strongholds in the hills where it was driven by the priority given in the past to agriculture . . . to offer a much wider range of forests and woodlands, and a much wider range of public benefits'.

So the agriculture authorities had at last softened their no-go approach to tree planting on better land, first introduced after the Second World War, in the face of food surpluses being generated by the EU's Common Agricultural Policy (CAP).

By 1980 the sheer volume of wood arising from the afforestation programme started to attract the interest of timber companies producing paper and wood panels. The early failure of the Fort William pulp mill had not undermined the case for industrial development, rather the contrary. Wiggins Teape had drawn attention to the opportunities that an uncommitted and growing wood resource in Scotland provided. It was the kind of industry in which only the established international companies with their technical know-how and marketing clout could hope to make a success. These were encouraging times for forestry interests that could at last see the prospect of a financial return on their woods.

Built by the Finnish company Kymmene, Caledonian Paper at Irvine was greeted in 1987 as Scotland's biggest ever inward investment. Highland Forest Products at Dalcross, when it started in 1985, was from the United States. Caberboard set up in 1982 on the site of the old Scotboard factory at Cowie and was a German investment. In contrast sawmilling remained the province of the established Scottish companies, some of whom we have already encountered. Through expansion and rationalisation, they sustained their trading activities almost uninterrupted throughout the whole century.

All this investment heralded a further phase of productivity increases in the woods. Purpose-built forwarders were far quicker and more agile than the skidders had been. Using the latest chainsaws still involved considerable physical effort and exposure to noise and vibration, so the forwarders were fitted with special heads for felling and snedding. In the right conditions, the new machines could now do the whole job – converting the standing tree to a stack of roundwood on a lorry – mechanisation that multiplied one man's output by a further factor of ten (see Colour Plate 12). It was progress of a sort, even if it frustrated the vision of those pioneers who had expected afforestation to repopulate the Highlands.

Changed Times – Multi-purpose Forestry

Society today has such an eclectic view on the value of woods that we need to remind ourselves that the nineteenth century's legacy was largely a utilitarian one, at least in respect of the ordinary landowners who had no money to invest in altruistic improvements on the land.

That utilitarian view of the world was given a fresh energy by the strategic imperatives that flowed from the two world wars and it was not until the 1970s and 1980s that a serious crossing of interests occurred. On the one hand, the post-war breakthroughs in forestry – particularly the mechanisation of afforestation – encouraged a degree of commercialisation that had not previously been seen. On the other hand, society's increasing urbanisation and wealth began to change the way people used and thought about the countryside and the woods within it and, in the 1980s, began to influence forest policy away from its preoccupations with wood production which had really been its *raison d'être* since 1919.

PLATE 7.18 *Tree-felling at South Laggan. 'Thinning' opens up the canopy to light and gives more room for the best stems to grow. Here, in 1961, cross-cut saws were still in use, though being progressively replaced by the chainsaw. Today, two men at a tree would be a breach of safety rules, as would the absence of hard hats and protective clothing. (Image courtesy of the Forest Research Library.)*

But that is to jump ahead. In the 1950s, Harry Steven, Professor of Forestry at Aberdeen University, was, along with a colleague Jock Carlisle, carrying out his classic study on *The Native Pinewoods of Scotland* (1959):

Even to walk through the larger of them gives one a better idea of what a primeval forest was like than can be got from any other woodland scene in

Britain. The trees range in age up to 300 years in some instances, and there are thus not very many generations between their earliest predecessors about 9000 years ago and those growing today; to stand in them is to feel the past.

If Steven allowed his romantic side to shine through his usually analytical descriptions, it was no more than most people felt when they visited the relict pinewoods, shrunk over centuries by exploitation and neglect to around a mere 17,900 hectares in eighty-four separate remnants. He was, of course, by no means alone in championing the pinewoods; there had been earlier work of a more academic kind and the very best of the pinewoods were already being earmarked for protection as nature reserves. However, Steven saw the big picture. He realised that their ultimate value was not so much as a museum piece for scientific study, but as a centrepiece of forestry and environmental policy. He died in 1969 not knowing if anything would be done:

> Moreover, land is short for forestry and there is, therefore, pressure to use what is available to the most productive end. But is land the shortest of our resources and will the destruction or the neglect of such woodlands contribute to forestry?

If Steven were writing today, he might use the word 'biodiversity' to describe the biological resources that he so much valued, not just the trees themselves but the wildlife, plants and soil that make up the whole wood and the story they tell about their use and abuse by humanity.

Only a tiny fraction of Scotland's woods are of natural origin with a direct line of descent from the wildwood of prehistoric times, even if modified in recent centuries by human design. Scattered about in the grain of the countryside – many hundreds of individual woods – they collectively cover some 152,000 hectares (about 2 per cent of the land area) typically pinewoods and birchwoods in the cooler and more exposed eastern and central Highlands, and deciduous woodland in the milder and wetter west – oak, ash, birch, rowan, hazel, alder, elm and others in mixture reflecting the soil's moisture and nutrition.

There was a stuttering start to the protection of native woods following the creation of the Nature Conservancy in 1949 (in its present form, Scottish Natural Heritage (SNH)). One of the most important relict pinewoods, on the flanks of Ben Eighe on Loch Maree, was acquired for the nation in 1951 as part of the first-ever National Nature Reserve. The west-coast oakwoods, with their rare lichens, mosses and liverworts, were protected in the Sunart and Taynish NNRs; nearby in Glen Nant, a Forest Nature Reserve, the woods were home to 150 different mosses and a similar number of moths. The birchwoods at Craigellachie and Inverpolly were chosen as special for their own characteristic biodiversity.

There were also protective measures that, in the later years of the century, would be seen as counter-productive. The wartime felling controls, which, in 1951, were retained for the new purpose of protecting amenity (by requiring replanting), did not

PLATE 7.19 *Striding out – a group of backpackers take the short cut. About two million Scots visit a wood every year. This photograph was taken in 1979. (Forestry Commission.)*

go so far as ensuring that ancient woods were regenerated with their native flora. Quite the contrary in fact – it was the fashion of the times to use introduced trees for their commercial value. The landowners, whose woods had been requisitioned for the war effort, saw in this a moral purpose as well as a material one – it was the thing to do in the 'reconstruction' spirit of post-war Britain and there was the hope and expectation that they were contributing to the creation of a new industry.

By the 1980s, however, pressure for a much more inclusive forest policy was mounting. The old order was being challenged, mainly by nature-conservation interests. It was a decade both of conflict and consultation between the different interests. There were the disputes over afforestation already mentioned and (in 1986) a report from the Nature Conservancy Council (NCC) on *Nature Conservation and Afforestation in Britain*, which voiced the Council's concerns about tree planting in the uplands. Meantime in 1982, the Institute of Chartered Foresters' Loughborough symposium on *Broadleaves in Britain* triggered a gradual move away from the single-minded commercial practices of earlier decades to a more balanced forestry policy.

There were complementary moves within government, too. In the late 1980s, the Treasury was encouraging Departments to count environmental benefits as well as commercial ones in public policy appraisal, making it easier for the Forestry Commission to adjust its grant schemes in favour of environmental outputs. Now it seemed that Steven's question in 1969 might get an answer. In 1988 the Forestry Commission's new Woodland Grant Scheme encouraged landowners to restore their native woods and create new ones. The Commission started work on a restoration programme in its own woods.

Creating and managing woods are costly and their owners seldom get a monetary return for providing what most people enjoy and think of as public benefits – landscape features or public access for instance. By the late 1980s, financial assistance from public funds was being seen as compensation for direct public benefits provided, rather than support for a strategic national purpose as the Dedication Scheme had been after the war. At the same time, woodlands started to attract some completely new sources of funding: grant-giving trusts, the National Lottery and even commercial interests sought to sponsor public-interest initiatives, such as safeguarding a threatened wood or recreational use.

In this, the funding agencies found common cause with land-owning environmental bodies. These organisations saw woodland ownership as a powerful means to their ends – ensuring that their objectives were met to the letter, giving them campaigning credibility as the owners and managers of their own woods and providing the opportunity for researching new approaches to practical woodland management problems.

So, in the late 1980s and 1990s, there developed a pattern of sponsored woodland acquisition and management by environmental charities. On the northern edge of the Lammermuirs, Woodhall Dene is one of the last remaining fragments of the ancient broadleaved forest of south Scotland, which the Scottish Wildlife Trust purchased in 1986 with help from the World Wide Fund for Nature. In the same year, the RSPB bought the main woods at Abernethy in upper Speyside with funding from the Nature Conservancy Council. Nearby, on the eastern fringes of the Cairngorms, the National Trust for Scotland set about restoring its newly acquired relict pinewoods at Mar Lodge in 1995 with the help of Forestry Commission grants. In the Trossachs, the Woodland Trust bought the 4,000-hectare estate of Glenfinglas in 1996 with a grant from the Heritage Lottery Fund, where they planned to return the degraded sheep walks to a mosaic of natural woods and pasture (see Plate 4.12).

It was not just nature conservation that benefited from the new funders. The 1980s saw the beginnings of an awakening interest in community woods, places where

PLATE 7.20 *Pony trekking in a forest park, 1979: walking and riding remain very popular activities in many Forestry Commission woods. (Forestry Commission.)*

recreation and nature could go hand in hand with the pleasure of shared ownership and, for those so inclined, manual work in woods – a wellington boots job (see Colour Plate 14). In 1979 the Countryside Commission for Scotland (whose functions are now embraced within SNH) started to encourage new community woods in the central belt as part of an environmental initiative to alleviate the area's legacy of urban and economic decline. The Central Scotland Countryside Trust (CSCT), which, in the 1990s, took over the initiative, received financial backing from a partner-ship of public agencies, tapping monies for its project work from the European Regional Development Fund and from the Landfill Tax, the first example of green taxation used to benefit woodland conservation.

Yet, curiously, it was the National Lottery that really put community woods on the map. A whole family of woodland-owning trusts was created through its Heritage Lottery Fund and the Millennium Commission. The Millennium Forest for Scotland Trust assisted community groups throughout Scotland with 'capacity-building' and funding help for new woods and the purchase of existing ones, as for instance the Abriachan Forest Trust on Loch Ness-side, the Cree Valley Community Trust in Galloway and the Borders Forest Trust.

Ownership of woods was nevertheless a step too far for most communities that merely sought a say in the *management* of their local woods. That the idea of community woods should catch on at all was in some ways a reflection of the way conventional forestry work – which after all started as little more than a craft industry – had become increasingly distant from the communities in which it had started and which no longer felt they had a stake in the benefits.

In truth there was little in the way of a *tradition* of forestry in most rural communities, but the forest workers, some of whom played a significant part in village life, nevertheless performed an largely unrecognised (if obvious) role as a conduit for information linking the community with forest activity. By the 1990s they had simply disappeared as a consequence of the slimming down of the forestry workforce referred to earlier, often coupled with the use of contractors that lived away from the rural areas.

It was a trend of employment that effected the traditional estates too, if perhaps to a lesser degree. They too had their perspectives on multi-purpose forestry, some seeing it as a return to the values of integrated land use that had long been the cornerstone of their particular style of woodland ownership. At Buccleuch, well known for its open-to-the-public estate woods, and for championing private estate forestry, the Earl of Dalkeith (reported in *Scottish Forestry* in 1995) summed up the feelings of many woodland owners:

> The emphasis, as we know, is no longer purely on output and production, but on the multi-purpose forestry that responds to demands for amenity and recreation requirements, the better understanding and aspirations for nature conservation and landscape improvement. For an estate with long-term horizons like Bowhill, it is not such a shift of the pendulum. In many ways it is just a move back to the more traditional objectives, and a welcome one at that.

There is also a sequel to our earlier account of wood utilisation. We saw how, after the Second World War, many of the 'traditional' sawmillers, who had until then worked with a mixed economy of tree species, began to concentrate on the new softwood resources that were expanding rapidly, thanks to afforestation. By contrast, some small companies nevertheless retained an interest in hardwood sawmilling, making products from native timbers.

By 1941, when Brownlie's were felling the last oak in the Binning wood (see earlier), Scotland's hardwood trade was already in decline. No one had planted hardwood trees for near on a century. From the landowners' point of view they simply took too long to grow. The markets were in decline. Oak, the prince of constructional materials for its toughness and durability, had long since given way to softwood in the construction of ordinary houses. Redwood and Oregon pine were lighter to carry, easier to work, and a lot cheaper to buy – imported as they were from the Baltic and North America.

So it was really not surprising that most end-of-the-century markets for oak should be in restoration. Small quantities of suitably shaped trees could still be found for one-off high-value uses. The repair of fishing boats – until the 1960s and 1970s a significant market for Scots oak – was a case in point (see Plate 7.10). There was also the restoration of ancient buildings; in 1997, for instance, 350 trees were felled for the great oaken trusses used to re-create the ancient roof of the Great Hall of Stirling Castle (see Colour Plate 11).

By comparison, markets for 'low-grade' hardwoods – more typical of the wood harvested in neglected native woods – were hard to find. Birch was the biggest resource of all. In their time, the deep mines had been a bulk market for hardwood chocks and other products. When this market fizzled out in the post-war years, difficult-to-handle short lengths and odd trees simply ended up as firewood, prompting the Scottish Native Woods Campaign to complain that 'oak and birch that goes up in smoke as firewood could become parquet flooring that can last over a hundred years'.

Highland Birchwoods concentrated on just such markets. Set up in 1992 as a charity offering advice on the management and use of native woods, it developed outlets in the buildings industry such as cladding, flooring and joinery products. The same idea was behind the Woodschool at Ancrum, where they made fine furniture from wind-blown trees and other people's waste wood. The twist was to see the wood market not as the main reason for the existence of a wood, but as one way of helping to secure its financial future.

So the question was whether new markets could be developed to stimulate the protection and regeneration of this neglected resource. If markets for commodity softwoods could be successfully developed from scratch, as they had been, could small-scale markets, and products, be developed for native hardwoods? Would the government's rural policies give greater support to small-scale rural enterprise? These were still open questions at the end of the century.

It was nevertheless with real expectation and authority that the Native Woodland Policy Forum, an alliance of non-government native woodland interests, was able to say in 1996:

> The promotion of native woodlands to a central role in Forestry Policy reflects important changes in society and should be seen as a fundamental step in the evolution of forestry in Scotland. It is a development which offers the prospect of renewing the vision of a national forest for Scotland.

LOOKING BACK AND LOOKING FORWARD

There was a remarkable change in the kinds of woods being planted in the closing years of the century when more than half the area planted and regenerated was of native trees, rather than introduced (mainly commercial) ones. This compared with less than 10 per cent in the 1970s at a time when commercial outputs were the main theme of public policy (see Fig. 8.3).

The attention given to statistics on tree planting and the regeneration of woods – figures which are collected and published every year by the Forestry Commission – surprises people outside the tree world who think it strange that forestry's health and welfare should be judged from *new woods* rather than *existing* ones. The problem of course is that no simple numbers can represent the condition of the nation's existing woodland estate, whereas new planting does indeed act as a weathervane on the mood of the country's woodland interests.

At the end of the twentieth century, it did seem that a lot was happening to manage *existing* woods of all kinds in ways that should provide public benefits. Over a hundred years, Scotland's stock of woods had grown from less than 5 per cent of the land surface to nearly 17 per cent – a dramatic transformation of the landscape, most noticeable in the uplands. Commercial woods were being 'restructured' to bring their ecological and aesthetic standards into line with the best practice embodied in a 'woodland assurance' code agreed between conservation and commercial interests. Real effort, at last, was going into the restoration of the ancient woods before it was too late. Furthermore, if in area terms the new community woods were tiny compared with other kinds of ownership, the numbers of people involved in them in the closing years of the century was anything but small, as was their energy and enthusiasm.

It is no criticism of the wood-using industry to say that the century's investment in woods has come from *external* finance, not generated by the wood market itself. On private land, new tree planting really only started when publicly-funded grants were introduced after the Second World War, initially as part of the Dedication Scheme. For centuries, and increasingly in Victorian times, the nation had relied on imported timber for the bulk of its timber needs. It follows that home-produced timber was, of necessity, priced in line with supplies from abroad, most of it from the exploitation of natural forests or from managed woods sold in the British market far below its replacement cost. Timber prices were, at the end of the century, worse than ever and if, rationally, they no longer provided landowning interests with a convincing argument for woodland investment, nevertheless they were not yet so bad or so prolonged as to weaken the landowner's traditional commitment to production forestry.

At the end of the century, wood production remained an objective of government, for its employment (about 10,700 jobs in Scotland, throughout all parts of the forest economy) and other industrial benefits, albeit now set in the widened arena of a multiple-purpose policy which encouraged diversity and gave emphasis to social and environmental benefits. The Forestry Commission, which at one time was the only external source of money for woods, remained an important funding agency at the end of the century, albeit increasingly under pressure to support the sustainable management of *existing* woods (including its own) rather than funding *new* ones. But it will be obvious from our story how the multi-purpose policy approach stimulated a new range of external funders.

Few people in 1900 would have foreseen what was to come in the new century and, at the end of it, we can do no better when we look ahead. Perhaps our only clue is the nature of the funders. We have seen how finance effects the very nature of the woods, where they are and what they look like. The biodiversity movement that flowed from the 'Earth Summit' in 1992 brought the depleted status of Britain's native woods into the general public arena in a way not seen before and was, at the end of the century, proving to be a powerful stimulant to public giving (through environmental charities) and to a range of funding agencies from the Lottery to corporate sponsors.

And then there is the CAP. We saw how woodland planting on farms was, in the last decade of the century, encouraged under European policies intended to diversify agricultural production. However, this was on a small scale. The prospect of serious reform to strengthen the rural economy represents a tantalising funding opportunity

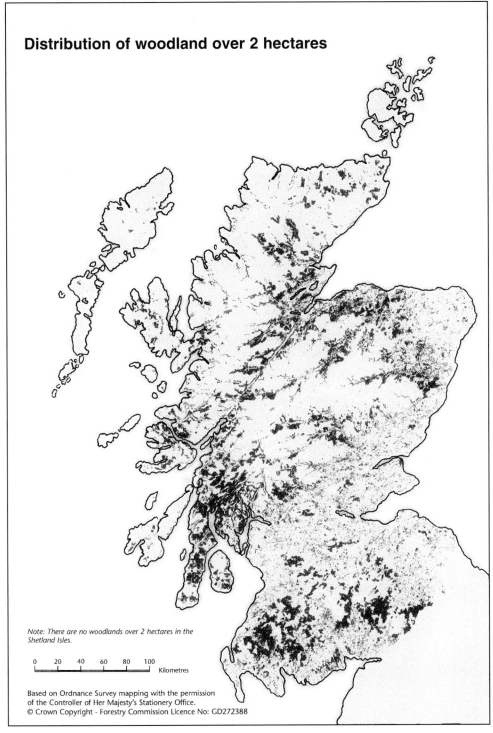

Distribution of woodland over 2 hectares

Note: There are no woodlands over 2 hectares in the
Shetland Isles.

0   20   40   60   80   100
                              Kilometres

Based on Ordnance Survey mapping with the permission
of the Controller of Her Majesty's Stationery Office.
© Crown Copyright - Forestry Commission Licence No: GD272388

FIGURE 7.1 *Map of all woodland in Scotland of over 2 hectares, at the end of the twentieth century. (Source: Forestry Commission Woodland Census, 1995.)*

for new woods and woodland restoration, and for small-scale wood-using enterprise.

Yet these themes are inter-locking in an era of multiple use. Woods for timber production can be established on farmland. Native woods produce timber and diversify wildlife of farms. Commercial forests provide recreational opportunities. Time will tell. Lord Tweedsmuir in his book *One Man's Happiness – A Countryman's Book* (1968) had it right:

> The planting of trees is the only thing in the world that gives you a clear reason for wanting to grow older. You look at a young plantation and wish that the next ten years would pass in a flash and your trees become a wood.

CHAPTER EIGHT

The Ecological Impact of
Using the Woods

RICHARD WORRELL and NEIL MACKENZIE

INTRODUCTION

The of use of woodlands has extended over such a long time period, and has been
so intense, that virtually no aspect of the ecology of woodlands in Scotland has
wholly escaped the influence of people. Whilst human activity has altered the ecology of woodland in most countries in the world, Scotland is exceptional in terms of the
huge scale of change wrought by people. These changes have been so profound that
there are very few relatively natural woodlands left. Whilst most native woodlands
may appear to be fairly natural, closer inspection reveals many ecological features
which result from past management. Furthermore, the rebuilding of forest cover during the twentieth century, primarily using tree species from North America, has left
us with a large part of our forest area designed and created by people. So whilst woodlands are often regarded as natural places, their ecology is, in fact, a fascinating mix of
things natural and artificial.

What are the main influences of people on native woodlands? These can be cate-
gorised as follows and each topic is described in detail in this chapter:

- Extent and distribution: how much woodland do we have left and how is it distributed? How has 'fragmentation' of woodland into small discrete woods affected
  its ecology?
- Tree species composition: how has the tree and shrub composition been altered
  by past use of woods?
- Woodland structure: how have the size and age of trees, their density and number
  of layers in the forest canopy been affected?
- Ecological processes: how have processes, such as tree regeneration, been influenced by people?
- Woodland animals and plants: how have populations of woodland animals and
  plants been changed, including extinctions of species such as bear, beaver and boar?
- Genetics of trees: how have people influenced the genetic make-up or 'genepool'
  of our woodlands?

PLATE 8.1 *(above, opposite and below). Birchwoods are the most widespread natural woodlands in Scotland, but they are often sparse, overgrazed by sheep and have little vegetation beneath them but coarse grass. (Neil Mackenzie.)*

It has often been assumed by ecologists that all the influences listed above are detrimental to native woodland, and indeed many of them have been. For this reason, examples of our remaining semi-natural woodland are regarded as highly valuable, particularly those which are relatively undisturbed and natural (e.g. remote gorge woodlands). However, some valuable native woodland habitats have been created by people's use of woodlands, for example wood pasture. It has been recognised only more recently that some of these 'artificial' woodlands can be very valuable ecologically, highly appealing and invaluable as living historical records. So we must not assume that all the results of the historical uses of woodland have reduced the ecological value of native woodlands.

A further set of questions arises out of the widespread planting of new woodlands, particularly conifer plantations. Many forests planted this century were established with little thought to ecological values. Conifer plantations have sometimes replaced native woodland or other valuable semi-natural habitat, and have accordingly drawn much criticism from environmentalists and the public. However, it is clear that these planted forests, given appropriate management, have also considerable potential to contribute positively to Scotland's ecology and biodiversity. We therefore need to establish the answers to questions such as:

- What is the ecological value of these forests and how do they compare with native woodland?
- What is the best balance to be struck between the future extent of planted conifer woodland and native woodland?
- How can these planted and native woodland be fitted together for maximum ecological value?

MODIFICATION OF SEMI-NATURAL WOODLANDS

Changes in Extent and Distribution

At the start of the twenty-first century we have a good understanding of how much native woodland remains in Scotland. Surveys of semi-natural woodland, aerial photography, accurate mapping and use of Geographical Information Systems (GIS) are allowing us to construct a comprehensive picture of the extent of our woodlands and where they are located. The most recent surveys estimate that 17 per cent of Scotland's land area is forested (19 per cent, if the Northern and Western Isles are excluded). Of that, 2 per cent consists of semi-natural woodland and 2 per cent of planted native forest (see Table 8.1). The great majority of our contemporary forests thus consist of non-native conifer plantations. Scotland contains a smaller proportion of native forests than any other country in Europe except Ireland.

So how has the extent of native woodlands changed during the past 100 years? We know from local studies and comparisons with Ordnance Survey maps dating from the early 1900s, that there was much more semi-natural woodland before the First World War than there is now (perhaps 3 to 4 per cent of the land area). We also know that the majority of planted forests at the start of the twentieth century were made up of native species, principally Scots pine and oak, although the non-native larch and Norway spruce were becoming increasingly important. During the twentieth century the proportion of native species declined due to:

- felling of mature woodlands, including ancient pinewoods during the world wars.
- the large-scale planting of non-native species for timber production, first by the Forestry Commission and then by private companies.
- the conversion of semi-natural woods and plantations of native species to non-native plantations. Birch and oak were extensively ring-barked and poisoned

and it is still possible today to see dead oak or birch standing among conifer plantations.

- Clearance of semi-natural woodlands for livestock grazing during the post-war era and further loss of these woodlands due to overgrazing by deer and sheep.

TABLE 8.1 The Woodland Resource in the Highlands and Lowlands of Scotland in 1999

	Highlands	Lowlands	Total
Total forest area	1,061,688	314,625	1,376,313
% of land area	21	14	19
Total conifers	849,128	265,686	1,114,814
% of land area	17	12	15
Total broadleaves	212,560	48,939	261,499
% of land area	4	2	4
Total semi-natural woodland	134,713	17,481	152,194
% of land area	3	1	2
Total other native woods*	149,231	19,513	168,744
% land area	3	1	2

* mainly planted Scots pine and some native broadleaves in the Highlands and planted oak, ash and elm in the Lowlands.
N.B. The land area calculated here excludes the Western Isles, Orkney and Shetland. Had they been included, the total forest area would have equated to 17 per cent of the land area, etc.
Source: N. A. MacKenzie, *The Native Woodland Resource of Scotland: A Review, 1993–1998*. Forestry Commission Technical Paper 30 (Edinburgh: 1999).

By 1950, native species in plantations had declined to about two-thirds of the forest area and by the 1980s the proportion was down to a quarter. At the same time the economic values formerly attributed to the native woodland were forgotten and the remaining woodland relicts were considered to have little importance. Birch was regarded by many foresters as a weed and birch regeneration in plantations was (and still is) routinely controlled. Only during recent years have moves been made to redress the damage done, by restoring native woodland on areas occupied by conifer plantations or open hill ground. Good examples include restoration of plantations on ancient woodland sites back to native woodland by Forest Enterprise in the pinewoods of Glenmore and the oakwoods of Sunart; and restoration of overgrazed woodland and hill land at Creag Meagaidh by Scottish Natural Heritage, where native woodlands are regenerating from lochshore to tree-line following the removal of sheep and a reduction in deer numbers.

The current distribution of native woodlands in Scotland reflects past history and land use. Regions where there are better agricultural soils, including much of the Lowlands, have long since lost their natural forest cover. As a result, semi-natural woodland in the Lowlands of Scotland accounts for less than 1 per cent of the land area. Relict woodlands still survive in river gorges, cleughs and a scattering of sites which have been managed for timber or shooting. A few new woodlands, notably

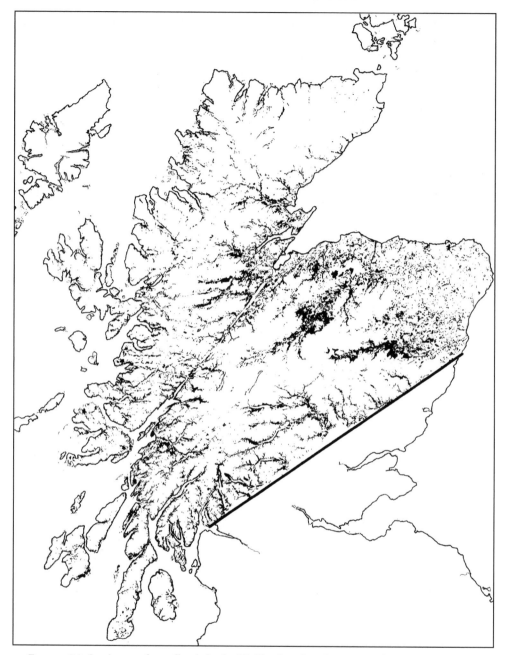

Figure 8.1 *Semi-natural woodlands in the Highlands today. Compare with Figure 7.1. (Copyright: Highland Birchwoods on behalf of Caledonian Partnership.)*

birch regeneration on derelict land or abandoned pasture, have established in recent decades.

The majority (almost 90 per cent) of Scotland's semi-natural woodlands are located within the Highlands, where they account for almost 3 per cent of the land area (see

Fig. 8.1). The soils are poorer, the human population density is smaller and many of the woodlands have survived because of their remoteness, because they were formerly managed for their timber or because there was no more profitable use of the land. Despite their small extent, native woodlands still form a significant component of the Highland landscape, particularly in well-wooded districts such as Argyll, Deeside and Strathspey. The birch- and pine-dominated areas of Cairngorms and the tributary glens of Strathglass contain some of the largest continuous tracts of native woodland now left in Britain (see Colour Plate 6).

Fragmentation and reduction in the size of woodlands have occurred universally, even in the well-wooded areas. Typically, native woodlands in the lower and middle levels of the glens have been reduced in area by forestry and agriculture, while grazing in the upper parts of glens has restricted the woodland fragments to steep slopes, rock ledges and scree (see Colour Plates 2 and 3). Apart from the one notable exception on Creag Fhiaclach above the Rothiemurchus pinewoods in the Cairngorms, tree-line woodlands have been lost throughout Scotland.

However, there have been episodes during the century when temporary reductions in grazing have permitted pulses of recovery. For example:

- the agricultural depression of the 1920s and 1930s allowed recovery and expansion of many birchwoods;
- reduction in the rabbit population for some years after 1954 due to myxomatosis led to an expansion of birch in many places;
- disturbance of deer by hillwalkers and skiers since the 1960s has allowed the pine and juniper to recolonise the northern slopes of the Cairngorms;
- changes in livestock or deer numbers or ranging patterns for a few seasons have permitted tree regeneration; and
- forestry fencing has permitted inadvertent regeneration of native broadleaves.

In the latter quarter of the century widespread concern at the loss of native wood-lands led to policy changes and financial incentives for the restoration of our natural forests. The last ten years in particular have seen dramatic increases in the regeneration and planting of native woodlands by both the public and the private sector. In the last five years of the millennium native species accounted for over half of all planting, restocking and regeneration under Forestry Commission grant schemes (see Fig. 8.3).

Ecological Effects of Fragmentation

An important effect of deforestation is the fragmentation of large forest areas into smaller patches of woodland isolated from each other by intervening agricultural land. Fragmentation appears to influence the presence and abundance of many woodland species, especially woodland birds and some insects. Fragments of woodland vary in their suitability as habitat – important factors including the overall woodland cover in the landscape, the size of patches of woodland, their shape (compact or elongated), and their degree of isolation. Effects of fragmentation are complex, but include:

- a reduction in habitat availability of some species which require larger areas of forest or forest interior conditions, especially woodland birds (e.g. capercaillie);
- an increase in abundance of some species which are adapted to woodland edge conditions (e.g. tits) or use both woodland and agricultural habitats (e.g. deer, fox);
- problems with pollination of some trees and shrubs due to isolation (e.g. aspen); and
- difficulties in migration and colonisation of some woodland species (especially plants) or movement of genes (see below) across the landscape.

It has been suggested that some of the negative impacts of fragmentation can be alleviated by joining woodland areas together again – a concept called the Forest Habitat Network.

Species Composition and Woodland Structure

Human activities over the centuries, including the last 100 years, have altered the species composition of most of our native woodlands. Birch is the most common type of semi-natural woodland in Scotland today (see Fig. 8.2). However, this has probably not always been the case, as birch is a successionary woodland type which becomes established immediately following a disturbance (though it is a natural climax woodland at higher altitudes and in exposed parts of the north-west). Birch is a common component of all types of woodland and the dominance of birch is often due to the removal of more useful species such as Scots pine, oak or ash for timber. A good example may be found on the limestone of Strath Suardal on Skye, where birch is now the only tree species present on a former ashwood on limestone pavement. Birch-dominated woods have also increased in extent because birch is efficient at colonising both woodland following felling and abandoned farmland. Similar effects can be found with other species; for example some pure hazelwoods on the west coast are the result of the removal of oak.

In contrast, many oakwoods and pinewoods, whilst appearing to be natural, have had the non-useful species (holly, cherry, willow, birch etc.) removed to favour the preferred timber trees and are therefore actually the product of selective management. The Sunart Oakwoods are a good example of this. This has frequently been overlooked by people wishing to conserve these woods, who have tended to regard their composition as natural.

The role of grazing has also been important in changing woodland structure and species composition. Overgrazing by sheep and deer has the following effects:

- Lack of regeneration: many native woodlands have not been able to regenerate for a very long time. This has resulted in the creation of an uniform age structure and limited species composition among the majority of woodlands. Many of the old pinewoods, such as those at Mar or Ballochbuie, have not regenerated for over two centuries because of the high numbers of red deer browsing on pine seedlings. Most of the trees in these woods are very old (up to 400 years) and near the end of their lives.

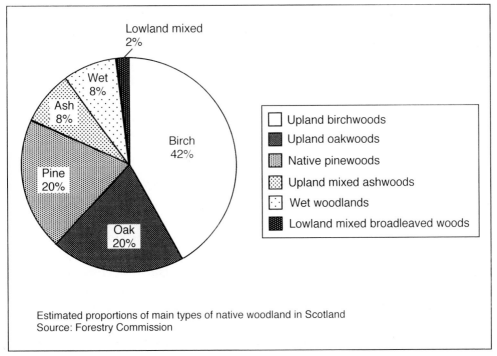

Lowland mixed
2%

Wet
8%

Ash
8%

Pine
20%

Birch
42%

Oak
20%

- [ ] Upland birchwoods
- [■] Upland oakwoods
- [▦] Native pinewoods
- [▨] Upland mixed ashwoods
- [⋰] Wet woodlands
- [■] Lowland mixed broadleaved woods

Estimated proportions of main types of native woodland in Scotland
Source: Forestry Commission

FIGURE 8.2 *Estimated proportions of the main types of native woodland in Scotland. (Forestry Commission.)*

- Elimination of tree species susceptible to browsing: aspen, willow, cherry and rowan are the preferred food plants of deer and livestock and tend to get eliminated from woodland. This is particularly striking in the case of pinewoods, which used to have a higher component of broadleaves.
- Reduction of shrub layer: the understorey of species such as holly, hazel or willow, and dwarf and creeping shrubs such as bramble, honeysuckle and wild rose are often highly vulnerable to browsing.
- Changes in the ground flora: high grazing levels tend to perpetuate a ground cover of sweet vernal, wavy hair, fescue and bent grasses among an abundant moss layer at the expense of tall herbs.

Most of these changes ultimately reduce biodiversity by removing the niches and food plants for a wide range of invertebrates, birds and mammals. However, some grazing is beneficial to woodland, and total exclusion of deer and livestock also leads to reduced diversity.

Wood Pasture

A particular feature of native woodland in Scotland are the remnant areas of wood pasture such as the pollarded alders and hazels of Glen Finglas and the ancient oaks of Cadzow (see Plates 3.7 and 4.12). These have particular ecological values because of the capacity of the ancient trees to support lichens, mosses and some types of invertebrate and birds. The importance of such habitats has only recently been recognised, and surveys are underway to determine the extent and characteristics of these types of woodland. Distinguishing between genuine wood pasture and overgrazed, moribund woodland, which has far less ecological value, is a particular challenge.

An interesting feature of parts of the genuine pinewood relicts is their parkland structure. Although these open grown pine have the appearance of wood pasture, this structure has often been developed from several centuries of burning, felling and grazing, rather than deliberate management as wood pasture. Examples can be found at Mar, Abernethy and Ballochbuie.

Woodland Dynamics and Ecological Processes

Woodlands change through time due to disturbance (e.g. wind, fire and disease) and the consequent process of succession, which is the sequence of colonisation by different tree and shrub species. Disturbance by people has become far more prominent, in addition to natural forms of disturbance. This includes felling of trees for timber and the setting of fires, either accidentally or on purpose (the latter sometimes happening with muirburn). For example, six major fires have occurred in Glentanar since 1688 (once every forty years), most thought to have been started by people. Disturbance cycles tend to be shorter now than in more natural forest. First, this is because of the dominance of species such as birch which have shorter disturbance cycles (60–100 years) and are not as long lived compared with, for example, oak. There are now relatively few examples of deciduous woodland on Lowland sites where natural disturbance cycles would often have been several hundred years. Second, many woodlands are kept artificially in an early stage of succession by timber management (e.g. pine plantations) and older woodlands are scarcely represented in the landscape at all.

Livestock farming and moorland management for sport have created large areas of pasture and upland heath where succession to woodland can potentially take place. The processes of succession have been changed by:

- reduction in the species diversity of woodland remnants and hence seed-sources for the colonisation of disturbed ground. This has contributed to the overrepresentation of birch.
- overgrazing by livestock and deer. This reduces the likelihood of palatable species surviving and again leads to reduced species diversity.
- the pasture vegetation itself, where it is more difficult for tree seedlings to establish themselves than in the vegetation usually found following forest disturbance.

Muirburn, the traditional management tool used to maintain upland pasture and heather for sheep, deer and grouse, has been shown to have both positive and negative influences on the age structure of the woodland. It can promote the creation of suitable ground conditions for natural regeneration but also often hinders regeneration by damaging established seedlings. In general, the negative impacts predominate and birch and pine regeneration is frequently destroyed by the fire.

In large natural forest areas a dynamic equilibrium can be established, with a variety of successional stages (from recent disturbance to climax forest) distributed throughout the landscape and changing their locations through time. This cannot happen in the current Scottish landscape of relatively small woodland blocks, where the land in between is not available for such a dynamic woodland cycle.

Changes to Woodland Soils

There is now considerable evidence that changes in soil structure and fertility have occurred as a consequence of deforestation. First, woodlands are effective at retaining nutrients compared with heavily grazed grassland, so loss of woodland is often accompanied by impoverishment of the soil. Transpiration and the interception of rainfall are reduced when trees are removed, which increases runoff and leaching and can contribute to soil waterlogging during winter. In heathy, upland areas an impermeable layer known as an iron pan can also develop, exacerbating waterlogging and peat formation.

Changes in Vegetation

Deforestation on upland sites often results in changes from a blaeberry–heather ground vegetation to the more peat-tolerant pure heather, purple moor grass or deer grass. Such habitat deterioration can lead to, or accelerate, blanket bog formation. Recent historical examples of this retrogressive succession are evident when early maps are examined and former woodland areas now appear to be bog or degraded wet heath (e.g. in parts of the native pinewoods of Strathfarrar or around Loch na Sealga, near Gruinard). Large expanses of moorland and grassland are easily perpetuated as the existing vegetation simply redevelops following disturbance because no other course is possible. However, some blanket and raised bog communities on deeper peats developed as a result of natural processes, aided by changes in the climate. Indeed, wetland, mire and grassland habitats are a natural component of any forest zone.

Deforestation has also created conditions for the opportunistic spread of bracken, which is a natural woodland fern, but has overrun grassland and heath to such an extent that tree regeneration can be restricted and diversity of plants reduced. Bracken can form dense, closed canopies up to two metres in height and become the dominant community. Muirburning and the change from cattle to sheep over much of the uplands have contributed to its expansion; the heavy feet of cows in the past broke up the rhizomes and inhibited its spread.

Woodland Fauna and Flora – Extinctions and Introductions

Mammals, birds and invertebrates are involved in many ecological processes in woodland including pollination, seed dispersal, nutrient cycling, the creation of seed beds and the maintenance of meadows, wetlands and flushes.

The extinction of a number of key forest mammals, including wolf, lynx, brown bear, beaver, wild boar, moose and auroch, occurred in earlier centuries as a result of hunting, persecution and habitat destruction. Other species, such as the polecat, red squirrel and capercaillie, became extinct or seriously reduced in number during the nineteenth and twentieth centuries. The osprey, goshawk and red kite disappeared but returned during the century by introduction or good fortune.

Deer, which, in the past, would have had their numbers and grazing behaviour modified by the predatory activities of wolf and lynx, are now solely controlled by culling. The loss of woodland has forced the red deer on to the more impoverished open moorland and this has dramatically altered their biology. Red deer living principally in a woodland habitat have been shown to achieve larger body sizes, higher fecundity and lower calf mortality than deer restricted to hill ground. The loss of the beaver has played a part in the reduction of diversity of both woodland and freshwater ecosystems, as these herbivores stimulated vegetation succession and increased the area of open space and wetland habitat. Wild boar, too, assisted regeneration and succession by their foraging activities, while aurochs, rather like cattle, probably selected and maintained open meadows in the forest.

Several species of mammal introduced deliberately during recent centuries continue to have an impact on native woodlands notably:

- Sika deer, which are hybridising with the native red deer;
- rabbits are locally a serious threat to tree regeneration;
- feral goats, which are escaped remnants of domestic herds, survive in numerous small pockets throughout the country and cause damage by browsing;
- grey squirrel is gradually spreading northwards into mixed and deciduous woodlands where it can outcompete the red squirrel.

Numerous trees and woodland plants have been introduced into Scotland, either for economic benefits or their ornamental qualities. A few of these have become invasive in native woodlands, notably sycamore, beech, rhododendron, some conifers and Japanese knotweed.

Sycamore was introduced several centuries ago and is now a common invader of mixed deciduous woodlands on fertile sites. The prolific numbers of seedlings are tolerant of shade and can outcompete most native trees. Recent research has shown that ash often regenerates under sycamore (and vice versa) suggesting that some sycamore/ashwoods may revert towards ash. Beech, although native to the south of England, can also colonise native woodlands on suitably fertile soils, especially gorge woods. Its canopy when mature casts such a dense shade that few vascular plants can survive under its canopy. Shade-tolerant softwood species such as the spruces, the firs and western hemlock are also most efficient colonisers of native woodland and in time can restrict regeneration of native species and limit the diversity and composition of the ground flora.

Nevertheless, sycamore, beech and non-native conifers are important timber trees and, on sites away from the semi-natural woods, can have important biodiversity functions. The base-rich bark of sycamore, for example, provides a valuable substrate for numerous lichens and old growth spruce forests can be an important habitat for fungi and some types of invertebrate such as beetles.

Rhododendron ponticum is a much more serious and insidious invader of native woodlands, becoming widely naturalised within oakwoods and acid heath communities, especially in the warm wet climate of the west of Scotland. The dense evergreen shade produced by the rhododendron and the toxicity of the roots not only prevent tree regeneration but eliminate all vegetation beneath its canopy.

Plant extinctions are more difficult to quantify because only in the last 100 years has there been sufficient botanical knowledge to record the rarer species. Since then, twenty-six species of plant have become extinct (not all woodland species) and many others are extremely rare as a result of habitat loss, collecting, ground disturbance and pollution. The beautiful little one-flowered wintergreen is now the rarest plant of the native pinewoods, having had its former distribution dramatically reduced by plant collectors and the felling of pinewoods. In Scotland no native tree species have become extinct but several, such as the montane willows and the rock whitebeam, are now very rare.

Genetic Changes

Genetics is a major determinant of many aspects of the lives of individual woodland species, including their growth, appearance, reproduction and, in the case of animals, behaviour. Some of the main aspects of ecology which are affected by genetics include:

- Adaptation: how well individual trees and shrubs are adapted to the site they are growing on. Different genetic varieties (called provenances) of tree and shrub species have evolved across Scotland which are specifically suited to the site conditions at the places they occur.
- Growth rhythms: the times at which different provenances of trees will flush (bud burst in spring), flower and cease growing in autumn are largely determined by genetics. This will in turn effect the other species dependent on the trees, especially pollinating insects.
- Tree form: whether trees grow straight or are twisted and branchy is partly determined by genetics. This is important when the trees are being managed for timber.

Historical and more recent use of woodlands has influenced their genetics in the following ways:

1. Fragmentation of the woodland area – leading to inbreeding (individuals mating with closely related individuals) and interruption of how genes 'flow' from woodland to woodland.
2. Changes in the woodland environment effect natural selection, i.e. which individual trees flourish and reproduce and which die.
3. Favouring or removal of particular types of tree during management – notably straight trees suitable for timber.
4. The planting of native trees grown from seed originating from foreign sources – which have different characteristics to Scottish ones.

Fragmentation of the forest area can lead to inbreeding and loss of genetic diversity. Inbreeding occurs when woodland stands have been reduced in size to fifty trees or less over several generations, and individual trees are then forced to mate with their relatives. This may lead to growth-rate reductions of up to 50 per cent in their offspring and other problems. For example, in the Glen Falloch pinewood remnant very low seed viability is evident and there is a high frequency of mutants in seedlings. Another potential problem in small isolated stands is that 'genetic variation' can slowly get lost (i.e. the genepool becomes progressively more narrow from one generation to the next). However, research has shown that even in woods like Glen Falloch, this does not appear to have happened to any significant extent yet, suggesting that the reduction to less than fifty individuals has only taken place in the last one or two generations of trees. This means that most remnant woodlands have not been significantly genetically impoverished and therefore can be used as seed-sources for future native woodland expansion.

A further effect of fragmentation is that it is harder for genes to move from woodland to woodland – which happens when pollen gets blown by the wind from wood to wood or when seed is dispersed. This process (called geneflow) is important in allowing populations of woodland species to adapt to changing climate by incorporating genes from distant woodlands where the trees are adapted to different site conditions. At present little is known about the extent to which geneflow is impeded by fragmentation, but it does potentially have serious effects on the capacity of woodlands to withstand climatic change.

Fragmentation also leads to woodlands being restricted to particular site types (often poorer quality or steep or inaccessible land). This means that our existing woodland resource has a disproportionately high number of trees adapted to these kinds of sites (as opposed to trees genetically suited to more fertile soils).

Changes in the woodland environment due to use by people have knock-on effects on natural selection in trees and shrubs. For example in a forest situation, trees that survive are those which are able to grow upwards fast and withstand shade and so compete with surrounding trees. In wood pasture quite different traits are probably an advantage: for example being able to germinate in a grazed sward, withstand grazing as a sapling, spread out sideways to occupy a maximum area of the site and produce seed when extremely old. Hence trees in wood pasture or grazed woods might be genetically rather different from those in forests.

Selection of trees during timber management influences the genetic make-up of woodlands. Management for timber inevitably means that trees with tall straight stems are preferred. This has two effects dependent on the type of management regime:

1. In poorly managed woods the straight trees have often been removed leaving only the poorer individuals. If this has happened over many generations, the genepool may contain a disproportionately high ratio of poorly formed trees, an effect called dysgenic selection.
2. In woods carefully managed for timber production, tall straight trees will have

been favoured and the poorer ones progressively removed. This will have had the opposite effect on the genepool.

A further effect of past management is that where coppicing or pollarding has been the main form of management, only those trees best able to form coppice/pollard shoots will have survived, so there will be a high proportion of such individuals. There is currently no information on the extent of these types of effect of past management.

An important aspect of genetics is the choice of seed origin of planted trees, that is, where the seed was collected from. It has been known for at least a hundred years that trees grown from seed from Continental Europe have different characteristics than Scottish ones. For example, as early as 1868 it was noted that foreign-origin seed of Scots pine, whilst it grew more rapidly for the first summer in the nursery, subsequently showed high levels of damage and poor survival thereafter compared with Scottish origin seed. Similar effects have been shown more recently for:

- birch, where Scandinavian origin plants often die in Scotland due to spring frost;
- oak, where French origin plants, together with ones from central and eastern European often fail or grow poorly – though some Dutch sources seem to survive adequately.

Seeds of native trees have been imported into Scotland from Continental Europe for several centuries, but it appears that use of these was very restricted compared with Scottish origin seed until the last thirty years – so we have few problems from these foreign imports except in recent woodlands.

However, seeds have also been imported from England, and, in the case of oak, northward transfer of acorns is thought to have occurred quite extensively. Experimental evidence suggests that oak of English origin survives rather less well than Scottish trees, but those which do survive generally grow adequately. Historic transfers of seed within Scotland also include frequent examples of seed being supplied from the Highlands for planting woods in Lowland areas – a trend which has in fact been reinstated in recent years.

The main effect of transfer of seeds and plants is that, in recent decades, the expansion of broadleaf planting has too often relied on imported seed, giving rise to plants which are poorly adapted to growing conditions in Scotland. This has resulted in the death of trees and poor growth. Fortunately, during the last few years there are signs that this problem is diminishing again as more Scottish seed is being collected and used by forest nurseries.

Wider Effects on Biodiversity

Human activities that directly affect native woodlands can be detrimental to adjacent ecosystems. Freshwater rivers and lochs depend on a healthy and diverse riparian tree cover for nutrient inputs in the form of leaves and woody debris, and the roots of trees for stabilising riverbanks. Native woodland is frequently sparse or absent along rivers flowing through agricultural fields or pasture and on tributaries in upper catchment systems, leading to poorer freshwater habitats.

Woodland ecosystems are also influenced by wider changes in the environment caused by people. These include:

- Intensification of management on surrounding land, which reduces habitat for some woodland species and restricts the connections between remaining woodland fragments (e.g. hedges, scrub, riparian woodland and meadow).
- Pollution, which can affect, for example, tree health and lichen diversity.
- Climate change, which is increasingly influencing all natural habitats, including woodlands. If the climate becomes warmer and wetter, effects may include a rise in the natural tree-line and loss of marginal habitats such as some subalpine willows, and wooded bogs. These problems are compounded by the fragmented state of woodlands in Scotland, which hinders migration of woodland species to new locations.

Planted Forests

Planted forests account for approximately 89 per cent of the current forest area of Scotland. They include a range of types from policy woods, to the familiar plantations of introduced conifers, to broadleaf plantations and new native woodlands (see Appendix 3 for definitions of these terms). The most important component of planted woodland is plantations of introduced conifers, which comprise 69 per cent of the national forest area and 78 per cent of the area of planted woodland.

Such plantations have both ecological benefits and impacts and, dependent on circumstances, they can be either beneficial or detrimental for conservation. The ecological characteristics of non-native conifer plantations vary greatly according to:

- the tree species involved;
- the style of management;
- the conservation value of the site prior to afforestation.

Plantations of species such as larch and lodgepole pine, which have some similarities to Scottish native species, are ecologically somewhat similar to native woodland, especially as they grow older. The main factor that drives these species differences is the 'shade tolerance'. Larches and pines are similar to Scottish native trees in having low-shade tolerance, which means that they need high levels of light to grow and that they only cast light shade. This allows the development of a woodland ground flora and associated invertebrate communities, with some similarities to that encountered in native woodland. In contrast, spruces and firs are very shade tolerant and cast a heavy shade, such that little ground flora can usually develop. However, these shade-tolerant species allow the development of other ecological niches – for example for fungi, beetles, mites and hoverflies – and so they also make a contribution to biodiversity.

Monocultural stands, managed on short rotations, have little to recommend them ecologically, except where they replace even more intensive land use such as arable fields. Fortunately such management is becoming somewhat less prevalent and many beneficial changes to management have been instigated in the last fifteen years, including the retention of open space, the inclusion of a broadleaved component, maintaining existing semi-natural woods, avoiding conifer planting adjacent to rivers, streams and wetlands and the retention of deadwood. Going beyond this, conifer plantations that are managed using ecological principles, for example natural regeneration, encouraging a diverse structure and composition and allowing trees to be retained to and beyond maturity, can have many beneficial ecological attributes. Management aimed at diversifying woodland structure and avoiding clearfelling (called continuous cover forestry) is now starting to be promoted.

Conifer plantations established on sites with low conservation value, such as arable farmland, some types of grazing land or derelict industrial land, are clearly a net gain for conservation. In contrast plantations established on ancient woodland sites, semi-natural mires or herb-rich pastures are usually ecologically detrimental. In these cases there are usually considerable ecological gains to be made by restoring the original habitat, which has started to be done in the case of ancient woodland and mire.

RECENT TRENDS IN ECOLOGICAL RESTORATION AND USE OF WOODLANDS

Recent years have witnessed a dramatic resurgence of interest in native woodlands, so that native woodland management has started to be adopted as a central part of forestry in Scotland (see Fig. 8.3). Large areas of native woodland, many of which have been neglected for most of the twentieth century, have been restored and significant areas of new woodland created by planting and natural regeneration. This reflects recognition on the part of forest policy makers that native woodlands have many benefits and that resources need to be split more evenly between native woodland and conifer forests. At the same time in planted conifer forests there has been an upsurge in interest in forms of management which utilise or mimic natural processes. Some of the key themes which are currently being explored are:

- Forest Habitat Networks: the creation of 'Forest Habitat Networks' aimed at linking areas of woodland throughout the landscape. These are thought to be ecologically more valuable than isolated blocks, because woodland flora and fauna can then migrate through the landscape.
- Biodiversity: native woodland Habitat Action Plans have been drawn up under a procedure initiated by European Union legislation. This process includes for the first time targets for how much existing native woodland is to be restored and how much new woodland is to be created.
- Rural development forestry: there has been a resurgence of interest in social and economic benefits of native woodland. So, after a break of a century or so, native woodlands are once again being seen as part of our rural resources.

- Woodland grazing: the long history and importance of woodland grazing is being recognised, with greater appreciation of the value of wood pasture and grazed woodlands.
- Integrating native woodlands and introduced conifer plantations: a debate has been taking place as to how the future forest resource should be split between native woodlands and conifer plantations. Whilst some in the forest industry believe that too many resources have been put into native woodlands in recent years, others in forestry (and beyond) believe that there is still a huge historical disparity to be addressed, and that development of the roles of native woodland holds the key to successful forests in Scotland. A further topic of debate is how to achieve integration of the two types of forest – by mixing native and introduced species in novel types of forest.

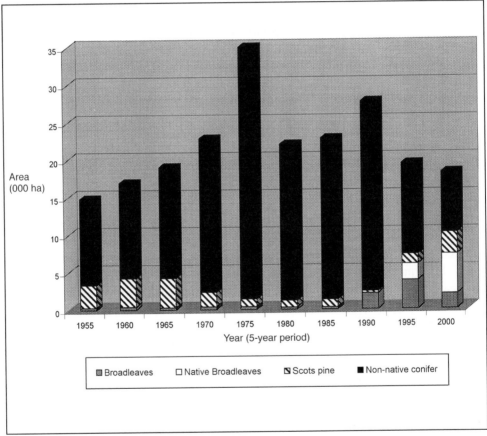

FIGURE 8.3 *Planting in Scotland since 1950: notice the revival of broadleaf planting since 1986. (Forestry Commission.)*

The trends above suggest exciting prospects for native woodlands and signal the development of a more holistic approach to forestry, which recognises the importance of both the productive use and the ecological values of woodland and forest.

Conclusion: Why is Understanding the Influence of Historical Use of Woodlands so Important?

Good management of woodlands in the future depends on understanding both the natural characteristics of woodlands and the characteristics, good and bad, that they have acquired as a result of their long interaction with people. Discerning between natural and artificial (historical) characteristics of native woodlands is not always easy and requires us to understand both ecology and history. Conservation management often aims to increase the natural attributes of woodland, in which case it is important to understand the influence of their past use in order to return the wood to a more natural state (e.g. diversify the species mix or structure). Sometimes the past use of woodlands has left them as valuable habitats and so we need to understand past use in order to maintain these values (e.g. wood pasture). There may also be instances where history can teach us about productive uses of woods which may be relevant today (e.g. coppice management). Realisation of the importance of this interface between ecology and history has grown rapidly in recent years and has broadened our understanding and appreciation of our woodlands.

CHAPTER NINE

The Future

ALEXANDER MATHER

INTRODUCTION

At the beginning of the twentieth century, who would have foreseen the tremendous changes in woods and forests that were to take place in Scotland over the next hundred years? At the beginning of the twenty-first century, to look into the future of 'people and woods' is a daunting task. Some of the trends that became apparent both in Scotland and in some other parts of the world at the end of the twentieth century offer some general indications, but not precise or detailed directions. What is clear is that the end of the twentieth century has been a turning point in how we perceive and relate to forests.

The next few decades are likely to see a continuing expansion of forests, a greater diversity of forests and woodlands and of the purposes for which they are used, and a closer relationship between people and woods in Scotland. In these respects, Scotland resembles many other parts of the developed world, where forests have expanded in recent decades and where the prevailing view of the forest has changed. The details of the local manifestation of these trends in Scotland are unique, but the trends themselves are widespread throughout the developed world.

HOW MUCH FOREST WILL WE HAVE?

The area of woods and forests in Scotland trebled during the twentieth century. If this rate of expansion were to continue in the twenty-first century, half the country would be under forest by 2100. Scotland would then be comparable to present-day Sweden or Russia in terms of its proportion of forest. This amount of expansion is unlikely, but the likelihood that expansion will continue over the next few decades is high. The *Scottish Forestry Strategy*, launched in 2000, talks of increasing Scotland's forests and woodlands towards one-quarter of the land area by the middle of this century. Currently, the proportion is approximately one-sixth. One reason for continuing expansion is that agriculture is likely to contract in area, and forestry is one of the relatively few alternative land uses. Another is simply that we seem to want more trees, woods and forests, according to the Forestry Commission's survey of *Public Opinion of Forestry* in 1999 (see Table 9.1). The clear message is that people in Scotland would like to have more woodland, but at the same time there is less enthusiasm for a dramatic increase. Most want only 'a little more', whereas in England the biggest single group of respondents opted for a doubling of the forest area.

TABLE 9.1 Would you like to have more or less woodland in this part of the country?

	(Percentage of respondents)		
	Scotland	England	Wales
About twice as much	17	30	19
Increase the area by half	19	25	23
A little more	41	28	37
Neither more nor less	12	12	12
Less	3	1	3
Don't know	8	4	7

In theory, the rate of transfer of land from agriculture to forestry could be greater than it was in much of the twentieth century. Mechanisms to protect agricultural land have been largely dismantled, and many farms are under severe financial pressure. Scares over BSE and foot and mouth disease have diverted our attention from long-term downward trends in consumption of beef and lamb. These trends do not make happy reading for farmers in the hills and uplands, whose range of choice of products is usually limited by inhospitable environments. Unless strong export markets can be found, the prospects for farming in many upland areas are gloomy, and conversion of land to forest is one option. The most likely scenario is active conversion through planting, but there is also the possibility that spontaneous natural regeneration could take place in some areas if sheep and their grazing pressures were removed. Much would depend on what happened to deer populations.

In terms of scenarios for forest expansion, Scotland resembles much of Europe. Land has been coming out of agriculture for several decades in many European countries, and is likely to continue to do so into the foreseeable future. Reforestation of released land would be welcomed by many, but of course it is not always an unmitigated blessing. In parts of the Alps, for example, forest regeneration on aban-doned pastures has resulted in landscape changes that are sometimes regarded as unwelcome. Also traditional commercial afforestation for purposes of timber pro-duction would not be favoured by all, especially if near-monocultures of exotic species were to be involved.

WHERE WILL THE NEW WOODS AND FORESTS BE LOCATED?

During the twentieth century, most afforestation was in the uplands, and much of it was concentrated in the south and west of Scotland (especially in Dumfries and Galloway and in Argyll). The upland emphasis reflected policy decisions to avoid planting on arable land, and had implications for choice of species and design of forests. Concentration was linked to economies of scale in management and to potential for supporting wood-using industry. From the mid-1980s onwards, this pattern began to change, as planting became less concentrated geographically, and somewhat more common in the Lowlands. This change resulted in part from the introduction of farm woodland schemes, after it was recognised that agricultural production had to be curbed and that farmers needed incentives if they were to become directly involved in forestry. It also resulted from initiatives such as the Central Scotland Forest, and the acceptance of the idea that woods and forests could serve a variety of purposes including landscape improvement in areas affected by mining and heavy industry. In short, the locations of new planting became more diverse, and this trend will probably continue into the foreseeable future. Small woods and forests will be established around the Lowland towns and cities, as pockets of land are transferred out of farming.

On the other hand, most of the larger tracts of land that are likely to pass out of farming are in the uplands and upland margins. Scotland has large areas of marginal land, often in relatively remote locations such as upland Moray. If such areas are afforested, recreational use is likely to be a lower priority than around the Lowland towns and cities, and greater emphasis will probably be placed on timber production. Grouse moors could be seriously affected by changes in fashion or growth of anti-sporting and animal-rights sentiments. If this occurred, huge areas of moorland could be available for afforestation. The question of whether such areas of marginal farmland and adjacent moors are ever afforested will depend strongly on economics, and in particular on the economics of timber production in other parts of the world. On the other hand, if grazing and browsing were restricted, natural regeneration could lead to scrub or forest in time.

What will the New Woods and Forests be Like?

Diversity will be a key theme, probably in terms of both species composition and design. The small woods around the towns and cities will probably be composed mainly of broadleaved species, while the large tracts of new forests in the uplands will have higher proportions of conifers. One of the most remarkable features of trends in forestry over the last decade of the twentieth century was the upsurge in broadleaved planting, from a negligible proportion in the early 1980s to around half by the late 1990s. In Scotland it is possible that enthusiasm for broadleaved planting is slightly lower than in England and is clearly lower than in Wales. Table 9.2, based on the Forestry Commission's survey of *Public Opinion of Forestry* in 1999, summarises the apparent attitudes. What is clear is that a 'mixed' species composition is strongly preferred by people in Scotland.

What will the New Woods and Forests be For?

Again, diversity will probably be a key theme. Most forests are likely to be multi-purpose, but the range of purposes and the relative emphasis placed on each purpose will vary from place to place. Around the Lowland towns and cities, the emphasis will be on recreation and environmental benefits. In the larger upland plantations, a greater priority will be given to timber production, but this will not mean that environmental objectives (for example) are ignored. One role of the forest that will be mentioned is that of carbon sink, even if in practice the contribution is quite limited. Forests planted since 1990 will, if planting continues at current rates, take up around 0.4 million tons of carbon annually by 2010, compared with annual UK emissions of around 150 million tons. As *The Scottish Forestry Strategy* indicates, forests will not be planted solely to store carbon.

TABLE 9.2 Would you prefer the new woodland to be conifer or broadleaved, or a mixture of the two?

| | (Percentage of respondents) | | |
	Scotland	*England*	*Wales*
Conifer	3	2	0
Broadleaved	21	28	42
Mixed	72	64	56
Makes no difference	4	6	2

In retrospect, the twentieth century may turn out to be anomalous – in both farming and in forestry – in terms of its emphasis on single-purpose management. Paradoxically, the closest we may get to single-purpose forestry could be in some areas of ancient woodland, where conservation of biodiversity may be the paramount goal in the same way as timber production was accorded primacy throughout most of the twentieth century. On the other hand, some of the conservation trusts that own such areas are keen to highlight recreational opportunities. Indeed, one of the features of the forest scene in Scotland around the end of the twentieth century was the growth of involvement of conservation trusts and other environmental groups, especially in relation to native woodlands. In some cases, further support was provided by industry (for example by BP-Amoco and Scottish Power). It will be interesting to see whether such growth is sustained in the years ahead, and to see how the long-term management of areas of native woodlands will evolve.

A CHANGING VIEW OF WOODS AND FORESTS

Major changes in our perception of woods and forests took place in the latter part of the twentieth century. This is true both of Scotland and of the wider world. As a sweeping generalisation, we stopped seeing forests simply as tree farms geared to the production of wood, and began to see them in much broader terms, as attractive environments that could offer a variety of goods and services. It is easier to chart this trend than to explain it, as the causes lie deep within societal and cultural trends.

Two hundred years ago, Europeans began to see forests as factories for industrial timber, whereas previously they had been viewed as a resource that could yield a range of goods, including food and drink, medicines, fodder and numerous other products and services. In essence, the changing worldview of the Enlightenment brought a changing view of the forest. Rationality and science could be applied to its management for the production of a scarce material – wood for industry. The managed forest with its ranks of uniform trees soon came to epitomise the order and regularity so cherished in the Age of Reason.

The idea of the forest as timber factory did not develop as fully in Scotland in the nineteenth century as in Germany and some other parts of mainland Europe. Most of our timber requirements were imported, and planting was very limited in extent. Most of the owners of woods and forests were interested in sport and landscape, and perhaps were less strongly motivated by economic concerns than some of their counterparts elsewhere in Europe. With the drive for increased home timber production after the First World War, however, the notion of the forest as timber factory became firmly established in Britain and in Scotland. It is true that during the 1930s, the Forestry Commission established national forest parks on some of their properties in scenic areas with high proportions of unplantable land. Nevertheless, for decades the overriding aim was to build up a strategic reserve of timber, and timber production was the primary function of the new forests and the driver of forest expansion. Near monocultures of exotic conifers, selected for their ability to produce timber from upland soils, became the norm. During the last ten to fifteen years of the twentieth century, this view of the forest was seriously challenged. Timber production is now often only one of a number of functions of the forest, alongside, for example, recreation and the provision of wildlife habitat. One striking symbol of this shift in Scotland is the felling to waste of timber plantations of non-native conifers in parts of the Highlands, where the emphasis is now on the creation of Caledonian pine reserves. Environmental goals have in such settings now replaced the industrial aim of maximising timber production.

The end of the twentieth century saw a similar challenge throughout much of mainland Europe and North America. In general terms, the primacy of timber production was questioned, and increasing priority was given to other forest functions. The challenge has been manifested in a variety of ways in different parts of the world. One of the most dramatic was the case of the northern spotted owl and the closure to logging of some of the federal forests in the Pacific North West of the United States so that its habitat could be protected. Another was the United Nations Conference on Environment and Development (UNCED) in Rio de Janeiro in 1992 and the resulting set of Forest Principles that emerged from the resolutions printed in the *Report*. While these were not legally binding on the parties to the conference, they indicate a degree of consensus that:

> Forest resources and forest lands should be sustainably managed to meet the social, economic, ecological, cultural and spiritual human needs of present and future generations.

Following UNCED, a meeting of European ministers with responsibilities for forest issues took place in Helsinki in 1993, and the ensuing *Declaration and Resolutions* in effect translated the Rio principles into guidelines for European forest management. Sustainable forest management and multi-purpose forests became normative:

> Forest management should provide, to the extent that it is economically and environmentally sound to do so, optimal combinations of goods and services to nations and to local populations. Multiple-use forestry should be promoted to achieve an appropriate balance between the various needs of society.

As a signatory, the United Kingdom subscribed to these guidelines, and translated them into British terms. A further European ministerial conference in Lisbon in 1998 focused on social- and community-related aspects of forest management, again with implications for the UK and for Scotland.

WHY HAS THE PREVAILING VIEW OF WOODS AND FORESTS SHIFTED?

Events such as those at Rio, Helsinki and Lisbon are milestones on the road towards forests that are more diverse in purpose and composition, but they do not in themselves explain why the trend has been in operation. The basic reasons are probably very deep-seated. One possibility is a basic trend in society in the affluent developed world. Satiated with material goods, people seek better lifestyle conditions and place greater emphasis on non-material values. An outworking of this is a swing away from primacy of timber production, and towards greater diversity, with positive contributions to landscape, environment and quality of life. Another – or related – possibility is a reaction against the scientific rationalism that the regularity of the traditional coniferous timber plantation of the twentieth century epitomised.

Whatever the underlying forces, there is little doubt about the extent and significance of the sea-change that occurred in the prevailing view of the forest in Scotland during the 1990s. This is not to say that the change was smooth or uncontested, nor that it is complete. Indeed disagreement was evident in a variety of arenas and not least in the consultation leading to the publication in 2000 of *The Scottish Forestry Strategy*. Argument, no doubt, will continue in the coming decades, as different points of view compete for ascendancy.

PEOPLE AND FORESTS IN SCOTLAND

In Scotland, people and forests have been separated for centuries. The early deforestation of most of the country has meant that the forest plays little part in folk memory, folklore and literature. A qualification is necessary in relation to trees. There was, for example, Wallace's oak, and more generally the rowan by the croft door has its own place in tradition, and was – and in some cases still is – not lightly removed.

Even the 'myth of the Caledonian Forest' – the idea that until fairly recent times much of Scotland was covered with dense forest – did not really pervade folk life: only a small part of the Scottish population was probably ever aware of it. It is difficult to think of a Scottish equivalent of the 'royal oak' or of Sherwood Forest, for example. It is even more difficult to think of Scottish equivalents of the folktales of central Europe or Scandinavia. While it is perhaps true that trees and woods feature in Gaelic poetry and literature, they are less evident in Scottish literature. There may be the occasional mention such as that of Sir Walter Scott's 'shaggy woods', but the literature is relatively silent on forests and forestry, in comparison with farming and farmland (for example in Burns and Grassic Gibbon). Perhaps this is simply a reflection of the fact that the forest has not played a central role in Scottish life in recent centuries, and even the expansion that has occurred over the last hundred years has had little impact in terms of national consciousness and identity.

Until recently, people have generally been separated from the forest, geographically, emotionally and managerially. Until the twentieth century, significant areas of woodland were usually restricted to the large estates and to the vicinity of the big houses. The woods were the laird's, and not the people's. The general rural population had little involvement with the forest. In general, separation did not necessarily mean alienation: it simply meant that forests did not occur in the consciousness of most of the population. In general, Scotland escaped the negativity of the arson attacks on landowners' woods, of the kind that occurred in Ireland a hundred years ago, just as it avoided conflicts over control of the forest such as those of La Guerre des Desmoiselles in France in the 1830s and 1840s. Nevertheless, there was little scope for positive engagement. Furthermore, the landlord–tenant system of land tenure meant that the typical farmer was most unlikely to be involved in forestry, and indeed might have viewed the forest negatively as a reservoir of foxes and other 'vermin'.

For most of the twentieth century, convergence between forest and people was limited. Farming and forestry were often in competition for land, and for much of the expansionary phase that began with the setting up of the Forestry Commission in 1919, there was little local influence over afforestation or forest management. Control, capital and management were external and often remote, and labour was the only local input. Little change occurred even when the relative role of the Forestry Commission in new planting began to shrink. Its place was simply taken by rich investors with little connection with the communities in the areas where planting was taking place. Even the tax incentives that drove much of the planting as late as the 1980s militated against local involvement. In relation to forestry, high-rate tax payers with large incomes from the City or from the worlds of sport and entertainment were favoured more than local people with more modest incomes. Local landscapes could be transformed by afforestation, but local people often had little say in the decisions about whether it should proceed.

Towards the end of the twentieth century, notable changes began to occur in the relationship between people and forests in Scotland. One was the introduction of farm woodland schemes. After an understandably slow start, farmers have embraced these schemes in some numbers. In the past, tree planting was at best simply outside

the range of practical possibilities on most farms and at worst a threat to the survival
of the farm. Now it is a distinct possibility. Furthermore, only recently has serious
thought been given to the possibility of grazing by livestock in the forest: the assumption for a long time was of incompatibility.

In addition, some community involvement in local woodland management began
to develop during the 1990s, with forest-management partnerships being established
for some (state) forests, as at Laggan, and the sale of others to local communities, as at
Abriachan near Inverness. As yet, such examples are limited in number, but the trend
towards increasing community involvement, and to a greater role for forestry in rural
development, is nevertheless welcome. At the same time, a trend towards democrati-
sation of forestry can be discerned. Again, the trend is of limited strength, and some
issues such as the relationship between the Forestry Commission and the Scottish
Parliament continue to be potentially contentious. One symbol of increasing openness
is the posting of summaries of grant applications on the Forestry Commission
website: a decade ago such applications were shrouded in confidentiality. Another
expression is in the incorporation of Indicative Forestry Strategies into structure plans,
on which public consultation is carried out by local authorities. These examples may
only be straws in the wind, but they do indicate the general direction of trend. This
trend accords, of course, with the Lisbon Declaration, and again is not peculiar to
Scotland.

Reconciliation of people and forests in Scotland will not take place overnight. By
and large, people have been separated from forests and woodlands for centuries.
Separation has been geographical, in terms of physical distance, practical, with most
people having little to do with the forest, and emotional, with most people probably
thinking little of the forest. The gap is narrowing in each of these dimensions.
Physical proximity has increased and will probably continue to do as the forest area
increases. Recreational use has increased, and with it favourable perceptions of the
forest. All the trends – control, purpose and location – of the new forests that we
expect to be created in the next few decades point in the same direction.

PEOPLE AND FORESTS: *THE SCOTTISH FORESTRY STRATEGY*

The publication of *The Scottish Forestry Strategy* in 2000 was a milestone in the
evolution of the relationship between people and forests in Scotland. Even if its vision
is not realised, it is of symbolic significance. With the strategy being 'laid before the
Scottish Parliament by the Scottish ministers', a new measure of direct democratic
control was symbolised. As in several other spheres, the UK dimension has weakened
relative to those of Europe and of Scotland. General principles to which the UK government has signed up have come from the Helsinki and Lisbon Agreements, and EU
agri-environmental measures have provided support for farm afforestation. From the
Scottish level has come the *Strategy*, and from local-authority level some of the details
of its implementation. Arguably, therefore, detailed control has been moving closer to
the people. The notion of partnership between various bodies and groups is prominent
in discussion of the implementation of the *Strategy*, where once the assumption was

that the expert and authoritarian body – the Forestry Commission – would be wholly responsible.

The overarching theme of the *Strategy* is that of sustainability, as might be expected from the influences of Rio and Helsinki. As usual, sustainability is interpreted as the 'three-legged stool' with economic, environmental and social dimensions. Each of these dimensions has important implications for the relationship between people and forests in Scotland.

Compared with many of its European neighbours, Scotland enjoys the advantage of rapid tree growth and thus shorter cycles. Compared with some other countries, however, it is at a disadvantage. In parts of Chile, New Zealand and various tropical and subtropical parts of the world, growth rates can be higher and in some cases land and labour cheaper. Environmental standards are not uniform throughout the world, and in some countries the notion of the traditional single-purpose forest may still be accepted. In short, timber production may be cheaper outside Scotland, and large-scale investment will probably find its way to the areas offering the best returns. It is doubtful, therefore, whether the large areas of marginal land that could be released from agriculture in Scotland will be afforested solely for purposes of timber production – or at least for the production of bulk industrial timber. While it may be true that the UK remains heavily dependent on imports of forest products, it is also true that across Europe, annual growth of timber greatly exceeds annual removals or harvests.

This is not to say that the economic function of Scottish forests is insignificant. Potential timber production will double over the next fifteen years, as the forests established during the 1970s and 1980s mature. Thereafter it is likely to dip towards mid-century, reflecting the slower rates of planting in the 1990s. It is unlikely that large numbers of new jobs will be created in the forest: at present, around 5,200 jobs are provided directly by forestry and a further 5,500 by haulage, processing and related activities. Overall, the employment-creating record of forestry in the twentieth century was disappointing. Many of the hopes expressed around mid-century, for example, failed to materialise, and initiatives such as forest villages and forest holdings were less successful than hoped. Perhaps there is now more realism about the extent to which the forest can employ people directly. In the Forestry Commission's survey of *Public Opinion of Forestry* in 1999, only half of the respondents in Scotland thought that bringing jobs to rural areas was a good reason for supporting forestry with public money. Interestingly, timber production was seen as a good reason for government support by less than one-quarter of the respondents in Scotland. On the other hand, it is possible that the growth of small-scale woodworking industries, geared to high-value niche markets, could make useful contributions in some localities even if overall employment is modest.

On the whole, therefore, the direct involvement of people in forestry in Scotland, in terms of primary employment, is limited and is unlikely to increase much in the foreseeable future. If a closer relationship is to continue to develop, it will have to be in other ways. One could be through the continuing development of farm woodlands, and the diversification of agricultural employment to incorporate elements of woodland management. Another, of course, is through recreation. According to the Forestry Commission's survey of *Public Opinion of Forestry* in 1999, around 60 per cent of respondents in Scotland reported that they had visited forests or woodland for recreational purposes 'in the last few years'. According to another survey in 1998 and reported in the Forestry Commission's *Forestry Statistics*, however, two-thirds (66 per cent) said that that had not visited forests during 'last 12 months'. Most surveys also point to a predominance of persons from the higher social classes among woodland visitors. There is therefore scope for more visitors and for visitors with more diverse backgrounds. Clearly the location of forests has an important bearing on the extent to which they are used for recreation. Those in accessible urban fringes and in major tourist areas will have much greater recreational use than forests in some of the remoter areas, where marginal land may come out of agricultural use.

Perhaps an equally significant way in which people and forests can come closer is through the familiar landscapes of home. In much of Scotland, both in the central belt and in areas such as Buchan and Caithness, the landscape has typically been open if not bleak. Such landscapes of course have their own qualities, but the simple fact is that, visually, woods and forests have not played large parts in the everyday lives of many Scots. Perhaps we have a case of 'out of sight, out of mind', and perhaps this is one of the reasons for the relative absence of woods and forests from Scottish literature and Scottish identity. With more woods and forests increasing in number and extent throughout most of the country, this absence may gradually become less noticeable.

The *Strategy* talks of 'conserving natural heritage and improving the environment', and makes it clear that this goal applies particularly to the protection and expansion of native woodlands, to increasing the value of other woods and forests in terms of biodiversity, and to minimising negative environmental impacts. An amazing transformation has occurred in the prevailing perceptions of Caledonian pinewoods. Half a century ago, they were often regarded as simply as 'scrub': when timber production was the main theme, their value was discounted. Now the wheel has turned almost full circle: some of the main timber-producing species such as Sitka spruce and Lodgepole pine are widely disparaged, and Caledonian pinewoods are cherished. More generally, some resurgence occurred in the planting of Scots pine in the 1990s after it had fallen to negligible levels in the 1980s, and this trend is likely to continue. Overall, many of the forests planted during the twentieth century are likely to be 'restructured' during the twenty-first, with more sympathetic design, less uniform age structure and more diverse species composition. The new forests of the twenty-first century will have similar characteristics. This is likely to be the case even in forests where timber production remains the primary function. In general, many of the principles of what Americans call 'forest ecosystem management' will be applied, notably in terms of diversity of purposes and types of woodlands.

Superimposed on global trends are, of course, local circumstances, including the political climate. The notion of reforestation as a symbol of regeneration is a common one: it occurs, for example, in the *England Forest Strategy* just as it did in relation to national regeneration in nineteenth-century France. Association with national regeneration in Scotland tends to be muted and implicit, although it is perhaps not absent from the rhetoric of movements such as Reforesting Scotland.

The vision of *The Scottish Forestry Strategy* is 'that Scotland will be renowned as a land of fine trees, wood and forests which strengthen the economy, which enrich the natural environment and which people enjoy and value'. At the beginning of the twentieth century, such a vision would have completely lacked credibility. Scotland might have had some fine trees and a few fine woods, but they were to be found mainly on the estates of a few rich landowners. Fine forests were as rare as might be expected in one of the world's most deforested countries. A hundred years on, the vision is more credible. There has been significant change in each of the dimensions represented in the vision. The contribution to the economy is significant if not huge, the contribution to the natural environment is more positive, and more people enjoy and value forests. The role of woods and forests in the Scottish environment – physical, economic and social – has changed radically. In short, people and forests have come closer together, and in all probability will continue to do so in the foreseeable future.

Appendix 1a

Common Alder	*Alnus glutinosa*
Ash	*Fraxinus excelsior*
Aspen	*Populus tremula*
Downy Birch	*Betula pubescens*
Silver Birch	*Betula pendula*
Dwarf Birch	*Betula nana*
Blackthorn	*Prunus spinosa*
Bird Cherry	*Prunus padus*
Wild Cherry (Gean)	*Prunus avium*
Elder	*Sambucus nigra*
Wych Elm	*Ulmus glabra*
Hawthorn	*Crataegus monogyna*
Hazel	*Corylus avellana*
Holly	*Ilex aquifolium*
Juniper	*Juniperus communis*
Pedunculate Oak	*Quercus robur*
Sessile Oak	*Quercus petraea*
Scots Pine	*Pinus sylvestris*
Dog Rose	*Rosa canina*
Guelder Rose	*Viburnum opulus*
Rowan	*Sorbus aucuparia*
Rock Whitebeam	*Sorbus rupicola*
Whitebeam	*Sorbus pseudofennica*
Whitebeam	*Sorbus arranensis*
Willow, Goat	*Salix caprea*
Grey Willow	*Salix cinerea*
Eared Willow	*Salix aurita*
Woolly Willow	*Salix lanata*
Downy Willow	*Salix lapponum*
Tea-leaved Willow	*Salix phylicifolia*
Mountain Willow	*Salix arbuscula*
Whortle-leaved Willow	*Salix myrsinites*
Dark-leaved Willow	*Salix myrsinifolia*
Net-leaved Willow	*Salix reticulata*

Some of these species are rare or have a restricted natural distribution in Scotland (e.g. Elder, Guelder Rose and the Rock Whitebeam, *Sorbus rupicola*) while *Sorbus pseudofennica* and *S. arranensis* are found only on Arran and Scots pine only occurs naturally in the Highlands. Other small shrubs like gorse, broom, ivy, dwarf juniper and additional willow hybrids could also have been included in this list. The status of crab apple, *Malus sylvestris*, as a native Scottish tree is unclear. Most specimens are associated with settlement or old cultivation and few are recorded from native woodlands. The case for yew, *Taxus baccata*, is also unproven although there are small seminatural stands of yew in remote coastal areas of Argyll.

Appendix 1b

THE PRINCIPAL SPECIES USED IN FORESTRY IN SCOTLAND

Scots Pine	*Pinus sylvestris*
Corsican Pine	*Pinus nigra ssp. laricio*
Logepole Pine	*Pinus contorta*
Sitka Spruce	*Picea sitchensis*
Norway Spruce	*Picea abies*
European Larch	*Larix decidua*
Japanese Larch	*Larix kaempferi*
Douglas Fir	*Pseudotsuga menziesii*
Grand Fir	*Abies grandis*
Noble Fir	*Abies procera*
Western Hemlock	*Tsuga heterophylla*
Western Red Cedar	*Thuja plicata*
Sessile Oak	*Quercus petraea*
Pedunculate Oak	*Quercus robur*
Beech	*Fagus sylvatica*
Ash	*Fraxinus excelsior*
Sycamore	*Acer pseudoplatanus*
Wych Elm	*Ulmus glabra*
Sweet Chestnut	*Castanea sativa*
Silver Birch	*Betula pendula*
Downy Birch	*Betula pubescens*
Aspen	*Populus tremula*
Poplar Cultivars	*Populus deltoides* *P. nigra* *P. trichocarpa*
Common Alder	*Alnus glutinosa*
Small-leaved Lime	*Tilia cordata*
Gean, Wild Cherry	*Prunus avium*

Source: Based on G. Pyatt, D. Ray and J. Fletcher, 'An ecological site classification for forestry in Great Britain', *Forestry Commission Bulletin*, 124 (2001).

Appendix 2

FORESTRY COMMISSION CLASSIFICATION OF NATIVE WOODLANDS

The Forestry Commission recognise six types of native, self-sown woodland, each of which possess a distinctive ecological and regional character. This simple classification is used in the Forestry Commission's *Forestry Practice Guides* to provide management advice for the different native woodland types in Scotland.

1. Lowland mixed broadleaved woods – a mixture of oak, ash, hazel, wych elm, alder and birch on fertile soils in the east.
2. Upland mixed ashwoods – dominated by ash and hazel and often with wych elm, oak, birch and rowan.
3. Upland oakwoods – dominated by oak, usually sessile, and/or birch. Found throughout uplands, especially in the west and often on acid soils.
4. Upland birchwoods – dominated by birch with occasional other broadleaves like rowan or goat willow. On acid soils throughout the uplands.
5. Native pinewoods – dominated by Scots pine but often with birch and occasionally other broadleaves. On acid soils in the Highlands.
6. Wet woodlands – dominated by alder, willow, downy birch or a mixture of these. Found on wet soils, flushed.

Appendix 3

TERMS USED TO DESCRIBE WOODLAND AND FOREST

- **Ancient woodland:** woodland with a long recorded history of site occupancy, known to be present for at least 250 years and assumed to go back to prehistory.
- **Caledonian forest:** primarily used to describe the original prehistoric woodland cover, prior to significant disturbance by people or during Roman times. Caledonian pine forest is the term used to describe the native pinewoods.
- **Introduced, alien or exotic species:** a species which has been moved to a region or site by people.
- **Native forest:** extensive mosaics of native woodland and other types of land, where native woodland constitutes a high proportion of land area.
- **Native species:** species which arrived in a region by natural means rather than with the help of people and which have remained there ever since.
- **Native woodland:** a generic term describing woodland composed predominantly of tree and shrub species that are native in that locality.
- **Naturalised species:** an introduced species which is sufficiently adapted that it effectively fills an ecological niche and, for example, regenerates itself and supports some native flora and fauna.
- **New native woodland:** primarily used to describe recently planted native woodland designed to mimic semi-natural woodland; but sometimes also used to describe new woodland recently established by natural regeneration.
- **Plantation:** woodland created by planting, usually with a regular layout of trees and usually managed for timber production.
- **Plantation on ancient woodland site (PAWS):** a conifer plantation established on a site known once to carry ancient woodland.
- **Planted native woodland:** planted woodland composed of tree and shrub species that are native in that locality.
- **Semi-natural woodland:** native woodland that has regenerated naturally, generation after generation and has never been through a planted generation.

Appendix 4

AREA OF HIGH FOREST BY PRINCIPAL SPECIES

Species	Area ha	%
Scots Pine	135,828	15
Corsican Pine	2,200	0
Logepole Pine	121,060	13
Sitka Spruce	522,925	58
Norway Spruce	34,744	4
European Larch	8,622	1
Jap/Hybrid Larch	55,034	6
Douglas Fir	10,269	1
Other Conifers	5,496	1
Mixed Conifers	7,976	1
Total Conifers	904,154	100
Oak	20,215	11
Beech	8,610	5
Sycamore	10,200	5
Ash	4,763	3
Birch	75,996	40
Poplar	490	0
Sweet Chestnut	77	0
Elm	901	0
Other Broadleaves	15,556	8
Mixed Broadleaves	54,323	28
Total Broadleaves	191,131	100
Total – all species	1,095,285	

Source: Forestry Commission

Select Bibliography and
Guide to Further Reading

GENERAL

The following contain material either for the whole span of woodland history or for a good deal of it.

Three are older books, but classics:

M. L. Anderson, *A History of Scottish Forestry* (London: Nelson, 1967). Two volumes. Some of its general conclusions are now dated but the detail of its data is unmatched. Recently reprinted.

F. F. Darling and J. M. Boyd, *The Highlands and Islands* (London: Collins, 1964). Still unsurpassed as a survey of the ecology of the Highlands, and a major influence on the thinking on Highland land use, but not a reliable guide to the causes of woodland decline.

H. M. Steven and A. Carlisle, *The Native Pinewoods of Scotland* (Edinburgh: Oliver and Boyd, 1959). An imperishable account: the best book ever on Scottish woodland history. Reprinted.

See also:

C. Dickson and J. Dickson, *Plants and People in Ancient Scotland* (Stroud: Tempus, 2000).

M. L. Parry and T. R. Slater (eds), *The Making of the Scottish Countryside* (London: Croom Helm, 1980), especially Chapter 10 on mansion and policy woodland and Chapter 12 on commercial use of woodland.

T. C. Smout, *Nature Contested: Environmental History in Scotland and Northern England since 1600* (Edinburgh: Edinburgh University Press, 2000). Especially Chapter 2.

T. C. Smout (ed.), *Scottish Woodland History* (Edinburgh: Scottish Cultural Press, 1997). A book of essays.

R. Tipping, 'The form and fate of Scotland's woodlands', *Proceedings of the Society of Antiquaries of Scotland*, vol. 114 (1994), pp. 1–54. Covers the whole of prehistory.

An excellent discussion of what the natural woodland might have looked like, the ecological processes of woods, and a full consideration of the varied issues in the conservation and restoration of woods:

G. F. Peterken, *Natural Woodland, Ecology and Conservation in Northern Temperate Regions* (Cambridge: Cambridge University Press, 1996).

INTRODUCTION

Some of the cultural dimensions of trees and woodland are particularly indicated by these books:

K. Basford, *The Green Man* (Ipswich: Brewer, 1978, reissued in paperback 1998).

F. and G. Doel, *The Green Man in Britain* (Stroud: Tempus, 2001).

Hugh Fife, *Warriors and Guardians: Native Highland Trees* (Glendaruel: Argyll, 1994).

J. Hunter, *On the Other Side of Sorrow: Nature and People in the Scottish Highlands* (Edinburgh: Mainstream, 1995).

T. C. Smout, 'Trees as historic landscapes: Wallace's Oak to Reforesting Scotland', *Scottish Forestry*, 48 (1994), pp. 244–52.

CHAPTER 1

For a full understanding of the most important sources of data on prehistoric woodlands, this text is excellent:

H. J. B. Birks and H. H. Birks, *Quaternary Palaeoecology* (Cambridge: Cambridge University Press, 1980).

There are a number of recent books that explore themes in the inter-relations between early communities and their environments. The following are a good selection:

K. J. Edwards and I. B. M. Ralston (eds), *Scotland:. Environment and Archaeology, 8000 BC–AD 1000* (Chichester: Wiley, 1997) covers in detail both the palaeo-environmental evidence and archaeological records for Scotland.

J. G. Evans, *Land and Archaeology: Histories of the Human Environment in the British Isles* (Stroud: Tempus, 1999) is a more impressionistic but exciting new study.

The archaeological records covered in this chapter are explored in several very well-illustrated accounts for the general reader:

C. R. Wickham-Jones, *Scotland's First Settlers* (London: Batsford, 1994).

P. J. Ashmore, *Neolithic and Bronze Age Scotland* (London: Batsford, 1996).

R. Bewley, *Prehistoric Settlements* (London: Batsford, 1994).

M. Parker Pearson, *Bronze Age Britain* (London: Batsford, 1993).

CHAPTER 2

The first three items are of a general nature in respect to the Iron Age, and the remainder explore the interface of forest history and archaeology in more detail.

I. Armit (ed.), *Beyond the Brochs: Changing Perspectives on the Atlantic Scottish Iron Age* (Edinburgh: Edinburgh University Press, 1990).

I. Armit, *Celtic Scotland* (London: Batsford/Historic Scotland, 1997).

P. Dark, *The Environment of Britain in the First Millennium AD* (London: Duckworth, 2000).

W. S. Hanson, 'The organisation of Roman military timber-supply', *Britannia*, 9 (1978), pp. 293–305.

W. S. Hanson, 'Forest clearance and the Roman army', *Britannia*, 27 (1996), pp. 354–8.

W. S. Hanson and L. Macinnes, 'Forests, forts and fields: a discussion', *Scottish Archaeological Forum*, 12 (1981), pp. 98–113.

R. J. C. Mowat, *The Logboats of Scotland*, Oxbow Monograph, 68 (Oxford: 1996).

P. J. Reynolds, *Iron Age Farm: The Butser Experiment* (London: British Museum, 1979).

R. Sands, *Prehistoric woodworking: the analysis and interpretation of Bronze and Iron Age toolmarks: Wood in Archaeology*, 1 (London: University of London, Institute of Archaeology, 1997).

R. Tipping, 'Pollen analysis and the impact of Rome on native agriculture around Hadrian's Wall', in A. Gwilt and C. Haselgrove (eds), *Reconstructing Iron Age Societies*, Oxbow Monograph, 71 (Oxford: 1997), pp. 239–47.

CHAPTER 3

The uses of wood in medieval building can be traced in the following (some are specialist and technical):

B. E. Crawford and B. Ballin Smith, *The Biggings, Papa Stour, Shetland: The History and Archaeology of a Royal Norwegian Farm*, Society of Antiquaries Monograph Series, 15 (Edinburgh: 1999).

B. A. Crone, *The History of a Scottish Lowland Crannog: Excavations at Buiston, Ayrshire 1989–90*, STAR Monograph, 4 (Edinburgh: 2000).

B. A. Crone, 'Native tree-ring chronologies from some Scottish medieval burghs', *Medieval Archaeology*, 44 (2000), pp. 201–16.

C. Mills and A. Crone, 'Tree-ring evidence for the historic timber trade and woodland exploitation in Scotland', in V. Stravinskiene and R. Juknys (eds), *Dendrochronology and Environmental Trends* (Kaunas, Lithuania: Vytautas Magnus University, 1998).

I. Ralston, 'Pictish homes', in D. Henry (ed.), *The Worm, the Germ and the Thorn* (Balgavies: The Pinkfoot Press, 1997), pp. 19–34.

G. Stell, 'Urban building', in M. Lynch, M. Spearman and G. Stell (eds), *The Medieval Scottish Town* (Edinburgh: John Donald, 1988), pp. 60–80.

The historical evidence of woodland use and management may be followed up in:

J. Dowden (ed.), *The Chartulary of Lindores*, Scottish History Society, vol. 42 (1903), no. 111.

John M. Gilbert, *Hunting and Hunting Reserves in Medieval Scotland* (Edinburgh: John Donald, 1979).

C. Rogers (ed.), *Rental-book of the Cistercian Abbey of Coupar Angus* (London: 1879–90).

CHAPTERS 4 AND 5

For general background on rural life in the seventeenth to nineteenth centuries, see:

A. Bil, *The Shieling* (Edinburgh: John Donald, 1990).

R. A. Dodgshon, *Land and Society in Early Scotland* (Oxford: Oxford University Press, 1981).

R. A. Dodgshon, *From Chiefs to Landlords: Social Change in the Western Highlands and Islands, c. 1493–1820* (Edinburgh: Edinburgh University Press, 1998).

T. C. Smout, *A History of the Scottish People, 1560–1830* (London: Collins, 1969).

I. Whyte and K. Whyte, *The Changing Scottish Landscape, 1500–1800* (London: Routledge, 1991).

The following are especially about woods in the period:

G. A. Dixon, 'William Lorimer on forestry in the Central Highlands in the early 1760s', *Scottish Forestry*, 29 (1975), pp. 191–210.

G. A. Dixon, 'Forestry in Speyside in the 1760s', *Scottish Forestry*, 30 (1976), pp. 38–60.

E. Grant, *Abernethy Forest: Its People and its Past* (Nethy Bridge: Arkleton Trust, 1994).

J. M. Lindsay, 'The history of oak coppice in Scotland', *Scottish Forestry*, 29 (1975), pp. 87–95.

J. M. Lindsay, 'The iron industry in the Highlands: charcoal blast furnaces', *Scottish History Review*, 56 (1977), pp. 49–63.

D. Nairne, 'Notes on Highland woods, ancient and modern', *Transactions of the Gaelic Society of Inverness*, 7 (1890–2), pp. 355–63.

T. C. Smout, 'Some problems of timber supply in later seventeenth-century Scotland', *Scottish Forestry*, 14 (1960), pp. 3–13.

T. C. Smout and R. A. Lambert, *Rothiemurchus: Nature and People on a Highland Estate, 1500–2000* (Dalkeith: Scottish Cultural Press, 1999).

M. Stewart, *Loch Tay: Its Woods and its People* (Scottish Native Woods, Aberfeldy: 2001).

J. Dye et al., *The Sunart Oakwoods – A Report on their History and Archaeology* (n.p.: Sunart Oakwoods Research Group, 2001).

Finally, these provide snapshots of how people saw their own environment at the time:

E. Burt, *Letters from the North of Scotland*, 1754 (various editions, most recent Edinburgh: Birlinn, 1998).

I. C. Cunningham (ed.), *The Nation Survey'd: Timothy Pont's Maps of Scotland* (East Linton: John Tuckwell, 2001).

D. Henderson and J. H. Dickson (eds), *A Naturalist in the Highlands: James Robertson, his Life and Travels in Scotland, 1767–71* (Edinburgh: Scottish Academic Press, 1984).

A. Macleod (ed.), *The Songs of Duncan Ban Macintyre* (Edinburgh: Scottish Text Society, 1952).

A. Mitchell (ed.), *Geographical Collections Relating to Scotland made by Walter Macfarlane* (Edinburgh: Scottish History Society, 1st Series, 1906–8). Three volumes of topographical description of the seventeenth and eighteenth centuries.

T. Pennant, *A Tour in Scotland 1769* (various editions, most recent Edinburgh: Birlinn, 2000).

D. Wordsworth, *A Tour in Scotland in 1803* (various editions, more recent Edinburgh: Mercat Press, 1981).

CHAPTER 6

The following provide a good account of the early history of Scottish foresters and plant explorers, though it is also essential to consult M. L. Anderson, *A History of Scottish Forestry*, referred to in the 'General' section above.

F. R. S. Balfour, 'A history of conifers in Scotland and their discovery by Scotsmen', *Report of the 2nd International Conifer Conference: Conifers in Cultivation* (London: Royal Horticultural Society, 1932).

C. Dingwall, 'A History of ornamental planting in Scotland', in *Trees in the Landscape – Their History and Conservation* (London: Garden History Society, 1998).

Duke of Atholl (ed.), *Chronicles of Atholl and Tullibardine Families*, 1 (1908), p. 265.

Forestry Commission, *Scotland's Forest Heritage* (Edinburgh: HMSO, 1990).

S. G. House, 'The Famous Trees of Perthshire: The Heritage of the Great Plant Hunters of the 19th Century and their Introductions', *Scottish Woodlands History Discussion Group Notes*, 3 (1998), pp. 10–12.

T. Hunter, *Woods, Forests and Estates of Perthshire* (Perth: Henderson, Robertson and Hunter, 1883).

A. R. Macdonald, 'That valuable branch of the common good: the forestry plantation of eighteenth-century Perth', *Scottish Forestry*, 51 (1997), pp. 34–9.

A. Mitchell (ed.), *Conifers in the British Isles*, Forestry Commission booklet no. 33 (London: HMSO, 1975).

A. Mitchell et al. (eds), 'Champion trees in the British Isles', *Forestry Commission Technical Paper*, 7 (Edinburgh: HMSO, 1994).

A. L. Mitchell and S. House, *David Douglas: Explorer and Botanist* (London: Aurum Press, 1999).

I. Whyte, *Agriculture and Society in Seventeenth Century Scotland* (Edinburgh: John Donald, 1979), especially Chapter 5.

CHAPTER 7

The files of the journal *Scottish Forestry* are a particularly rich source for all aspects of Scottish forest history in the twentieth century.

G. Ryle, *Forest Service* (Newton Abbott: David and Charles, 1969) surveys the first half century of the Forestry Commission's existence.

W. Mutch, *Tall Trees and Small Woods* (Edinburgh: Mainstream, 1998) is a very readable book on modern forestry in Britain.

Two books by J. Sheail, *Rural Conservation in Inter-war Britain* (Oxford: Oxford University Press, 1981) and *Nature in Trust* (Glasgow: Blackie, 1976), put the work of the Commission into a wider context.

J. Tsouvalis, *A Critical Geography of Britain's State Forests* (Oxford: Oxford University Press, 2000) is a modern study of the Commission, though rather theoretical for the lay reader.

A number of books describe what it was like to work in the woods in the twentieth century:
J. Davies, *The Scottish Forester* (Edinburgh: Blackwoods, 1979).
A. Gray, *Timber!* (East Linton: Tuckwell Press, 1998).
J. McEwen, *A Life in Forestry* (Perth: Perth and Kinross Library, 1998).
N. Watson, *The Roots of BSW Timber plc* (Leyburn, North Yorks: St Mathew's Press, 1998).
W. C. Wonders, 'The Canadian Forestry Corps in Scotland during World War II', *Scottish Geographical Magazine*, 103 (1987), pp. 21–31.

CHAPTER 8

The ecological background can be further explored through the following:
J. R. Aldhous (ed.), *Our Pinewood Heritage*, Conference Proceedings (n.p.: Forestry Commission, Royal Society for the Protection of Birds and Scottish Natural Heritage, ISBN 0 88538 325 9: 1994).
R. Ennos, R. Worrell, P. Arkle and D. C. Malcolm, 'Genetic variation and conservation of British native trees and shrubs: current knowledge and policy implications', *Forestry Commission Technical Paper*, 31 (Edinburgh: 2000).
Highland Birchwoods, *Scotland's Semi-natural Woodland Inventory* (forthcoming).
R. A. Lambert (ed.), *Species History in Scotland* (Edinburgh: Scottish Cultural Press, 1998).
N. A. MacKenzie, 'The native woodland resource of Scotland, a review 1993–1998', *Forestry Commission Technical Paper*, 30 (Edinburgh: 1999).
G. F. Peterken, D. Baldock and A. Hampson, *A Forest Habitat Network for Scotland*, Scottish Natural Heritage Research, Survey and Monitoring Report, 44 (Battleby: 1995).
P. R. Quelch, *An Illustrated Guide to Ancient Wood Pasture in Scotland* (Glasgow: MFST Award Report. Millennium Forest for Scotland, 2001).
P. Ramsay, *Revival of the Land – Creag Meagaidh National Nature Reserve* (Battleby: Scottish Natural Heritage, n.d.).

CHAPTER 9

The official path for the future is laid out in:
Forests for Scotland: The Scottish Forestry Strategy (Edinburgh: Scottish Executive, 2000).

Also relevant are:
Rural Development Forestry (Battleby: Scottish Natural Heritage, 2001).
T. R. Lee, 'Perceptions, attitudes and preferences in forests and woodlands', *Forestry Commission Technical Paper*, 18 (Edinburgh: 2001).

List of Contributors

Ian Armit, Senior Lecturer, Department of Archaeology, Queen's University of Belfast.

Anne Crone, AOC Archaeology, Loanhead, Midlothian.

Christopher Dingwall, freelance landscape historian specialising in garden history, Washington House, Main Street, Ardler, Blairgowrie, Perthshire.

David Foot, former Forestry Commissioner and Head of the Forest Authority, 36 Coltbridge Terrace, Edinburgh.

Syd House, Conservator, Forestry Commission, Perth Conservancy, 14 Gowans Terrace, Perth.

Neil Mackenzie, ecological consultant, Norbu, Lochgarthside, Gorthleck, Inverness-shire.

Alexander Mather, Professor of Geography, Department of Geography and Environment, University of Aberdeen.

Ian Ralston, Professor of Archaeology, University of Edinburgh.

Christopher Smout, Centre for Environmental History and Policy, University of St Andrews.

Mairi Stewart, Prospect House, Home Street, Aberfeldy, Perthshire.

Richard Tipping, Senior Lecturer, Department of Environmental Science, University of Stirling.

Fiona Watson, Director, Centre for Environmental History and Policy, University of Stirling.

Richard Worrell, Woodland consultant, Upper Park, Aberfeldy, Perthshire.

Index

All arrangement is alphabetical except where there are a number of subheadings referring to periods of time. In this latter case these appear in chronological order and precede any other subheadings.